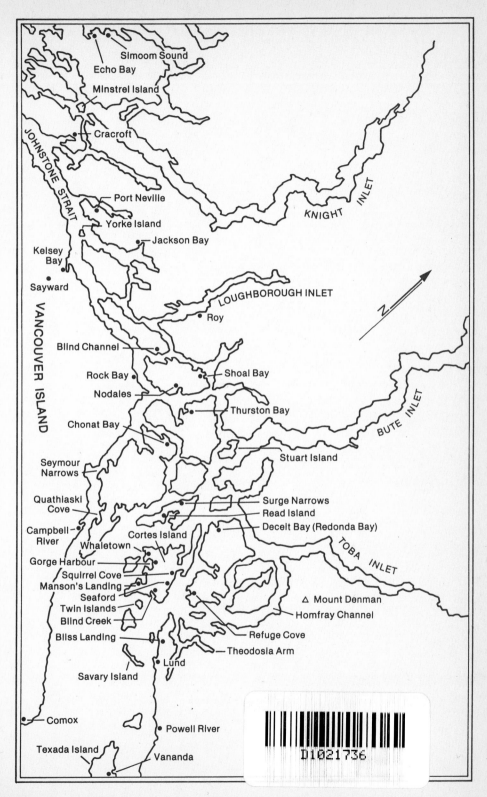

Simoom Sound
Echo Bay
Minstrel Island
Cracroft
JOHNSTONE STRAIT
Port Neville
Yorke Island
Jackson Bay
KNIGHT INLET
Kelsey Bay
Sayward
LOUGHBOROUGH INLET
Roy
VANCOUVER ISLAND
Blind Channel
Rock Bay
Nodales
Shoal Bay
Thurston Bay
BUTE INLET
Chonat Bay
Seymour Narrows
Stuart Island
Quathiaski Cove
Surge Narrows
Read Island
Deceit Bay (Redonda Bay)
Campbell River
Whaletown
Cortes Island
TOBA INLET
Gorge Harbour
Squirrel Cove
Manson's Landing
Seaford
Twin Islands
Blind Creek
△ Mount Denman
Homfray Channel
Bliss Landing
Refuge Cove
Theodosia Arm
Lund
Savary Island
Comox
Powell River
Texada Island
Vananda

Spilsbury's Coast

PIONEER YEARS IN THE WET WEST

Howard White
Jim Spilsbury

Harbour Publishing

Harbour Publishing Co. Ltd.
P.O. Box 219
Madeira Park, BC   Canada   V0N 2H0

Cover pastel by A.J. Spilsbury
Cover design by Gaye Hammond
Printed and bound in Canada by Gagné Printing

This book was written with the assistance of a Canada Council Arts
B Grant. Publication was financially assisted by the Government of
British Columbia through the British Columbia Heritage Trust.

CANADIAN CATALOGUING IN PUBLICATION DATA

White, Howard, 1945–
    Spilsbury's coast

    ISBN 1-55017-046-5
    Hardcover ISBN 0-920080-57-X

    Frontier and pioneer life—British Columbia. 2. British
Columbia—History. I. Spilsbury, Jim, 1905–   II. Title.
FC3817.3.W448 1987   971.1   C87-091394
F1087.W448 1987

Contents

This story is dedicated to the people in it; to all the loggers and stumpfarmers and hermits and fishermen and fisherwomen who knew Jim Spilsbury in those days, helped him on his way and made life along the coast so interesting. It is also dedicated to the many who directly worked with Jim in doing so many of the things for which he has inevitably received credit. To begin naming them would be to risk missing one, so we must rely on them to know this book is theirs with deepest gratitude.

Introduction

DURING THE FIFTIES when I was a boy growing up in my dad's logging camp on Nelson Island I used to think nothing we did quite counted. I owned a bit of a reputation around camp for the way I could skip across a slimy boomstick, but when I looked in the grade one reader my correspondence course provided, which I did almost monthly, the boys and girls there walked on sidewalks. All mention of boomsticks was carefully avoided. I could run the camp tender home from Garden Bay when the men got too drunk to do it themselves, but Dick and Jane could run elevators in apartment buildings, something I was sure I could never manage. Their fathers worked in offices, not under broke-down logging trucks.

Nothing was for us. The weather forecasts, on those rare occasions when our radio worked, always talked about "the lower mainland" and "the Kootenay region," never Green's Bay. We had weather in Green's Bay too, and a lot more of it out in Agamemnon Channel sometimes, but the world just sniggered at the thought of it, apparently.

Everything came from somewhere else. Even things that were specially ours, like Caterpillar tractors or ball bearings, came from places like Peoria Illinois or Kalamazoo Michigan. Those were the real places. The boys there would no doubt make short work of me. I seemed to be lost in a non-place inhabited by shadows, and I felt it deep inside.

There were only a couple of things that didn't fit the pattern. One was Easthope motors. They were made by people around the coast. Pete Dubois' uncle was married to one of the Easthope's sisters. But they were still from Vancouver, and Vancouver was the City, closer to Peoria than Pender Harbour.

More amazing was the Spilsbury and Tindall radiophone. We didn't actually have one of these in camp, but all the bigger fishboats did, and all the tugs and government boats. You'd see them in wheelhouses, jutting out from the wall, the sender and receiver separate, like two cases of Pacific milk. They had very fancy meters on them, banks of silver switches and dials, and wrinkle paint like binoculars. They looked as real and big-time as anything from Kalamazoo. Radio-telephones were even more impressive than ball bearings because they were more scientific and amazing. Yet they were made by a guy who lived up by Powell River, on a little island just like ours. He even had a boat, so I was told, and came around to the camps like Pappy's store boat. He'd never come into our camp, but a lot of the men knew him.

This filled me with wonder. If one guy around here could get onto the Kalamazoo level, it was at least possible. We weren't trapped here in limbo, just kind of bogged down, like the D8 the time they tried to drive across the cranberry bog at Goose Lake. Perhaps the reality we lived was not quite so counterfeit as I had come to think, if things like the S & T radiophone could come out of it. Spilsbury and Tindall became a very important symbol in my juvenile cosmos.

If I had known just how much Jim Spilsbury, the man who made those radio-telephones, was "a guy from around here," and how much of the unique edge-of-the-world culture of our coast was woven into his achievement, I would have been even more excited. When I became involved in writing down the story of the coast many years later I began to run across the Spilsbury name in a remarkable variety of contexts. Frank Lee, one of Pender Harbour's more senior pioneers, told me he had been neighbours with Spilsburys in Whonnock, before the Lee family moved to the coast, which seemed like before the dawn of time. The Spilsbury family were considered Whonnock's earliest settlers and have their name on a principal road, but when I began looking in on the history of the Powell River-Savary Island area, I found the name again on an important landmark, the northern tip of Hernando Island. That area claimed them as pioneers too. Lorne Maynard, an old coastal skipper, told me I should see Spilsbury about steam logging because he'd run steam donkeys. Geordie Tocher, who gained a certain amount of grudging respect among the oldtimers by sailing a BC fir log to Hawaii, told me if I wanted to see the most authentic paintings of coastal landscape ever done, I should go see Spilsbury's collection. When I started looking at the story of pioneer flying on the coast, everyone said Spilsbury was the man who had been in the middle of that.

It was to get the flying story that I finally did look Spilsbury up, and the tale he had was such an entertaining one I decided to make a book of it. But once I began probing the man's background I realized his life was far more than flying, and far more than radio. Eighty-two at the time of writing, his life has paralleled the history of the coast through this, its most active century. Because of his energies and his intelligence, he was involved in almost everything that happened. It is hard to name an erstwhile stumprancher or gyppo logger between Cape Caution and Point Atkinson that he doesn't have an anecdote about, usually a good one. It is hard to name an activity from homesteading to steam logging to police work to pleasure boating and mountain climbing that he wasn't personally deeply involved in. And all his memories are clear, backed by a vast collection of photos and journals, and told with all the warmth of a great personality.

Cold type isn't a wholly adequate medium in which to capture such a personality, particularly in the hands of one who never did finish his correspondence lessons, but in so far as I succeed, I succeed in showing the coast at its best.

HOWARD WHITE
Pender Harbour, 1987

ONE

Red's Sea Diner

WHEN I THINK OF THE WAY the coast was in the old days I think of Red Mahone. I don't know why. I didn't see as much of Red as I did a lot of people. I guess it was because Red was one guy I couldn't imagine happening anywhere else.

I first met Red when he was operating a restaurant (I use the word advisedly) in a floathouse at Minstrel Island, a small steamer stop at the mouth of Knight Inlet. The words *Red's Sea Diner* were displayed on a sign over the door. Minstrel Island was a busy port of call in those days and we would often eat at Red's when we blew in. There was nowhere else to go.

What was so typically BC coast about Red was the story of how he came there.

He had been a camp cook for many years, both in logging camps and on the towboats. He was a damn good cook by any standards, and he knew it. Red knew he could get away with anything he liked, and generally did.

After a good many years of terrorizing hungry loggers, Red decided to settle down with his wife in Campbell River and make his living operating this floating restaurant. They lived aboard. By virtue of a small outboard motor mounted in a well in the back of the scow, he could be self propelled and this gave him a wonderful feeling of independence. I just could not imagine Red being anything but independent, but it was not to last.

Frank Gagne was a contractor in Campbell River specializing in building and repairing docks. He had a pile driver and a small tug and everything that went along with the business, but he did not have a cook, or a cookhouse, and he had just landed a very juicy contract from the Federal Government to repair the wharf at Minstrel Island. Red and his portable restaurant would be just the ticket.

The only hitch was Red was much too in love with his new civilian life and told Frank to "bug off," or words to that effect. Frank persisted. Red told him he could take his piledriver and his crew and stick 'em.

It looked like a standoff, but Frank had an ace up his sleeve.

The day arrived when he was to move his show up the Straits. In order to catch the tide in Seymour Narrows they had to leave at midnight. Red and his wife were very sound asleep in their little floathouse. Frank's men, who didn't plan on eating out of cans for the next six months, did their work well.

When Red woke up in the morning to get ready for his early customers, he looked out and rubbed his eyes with disbelief. He was about fifty miles up Johnstone Straits tied alongside a piledriver behind a tug making five knots. When they arrived at Minstrel, Frank, remembering that Red was self-propelled, lowered a sixty-foot pile gently through the well in the stern of the houseboat and drove it six feet into the mud.

Eventually Frank and his well-fed crew finished the wharf job and moved away. Red was duly offered a free tow back to Campbell River, but he'd come to fancy life at Minstrel Island, and besides, that sixty-foot piling through the works gave him a wonderful feeling of security.

Once the deputy minister of transport came out from Ottawa to survey his western domain and I was honoured with the task of showing him around the upper coast. We took him to Minstrel Island because it was the most typical coastal logging community you could find, and we arrived just in time for lunch.

The scow the Sea Diner sat on was twenty feet long and ten wide. The house was eighteen feet long, leaving a sort of back porch just big enough for the engine well, which now had the creosote pile driven through it. The back half of the house was the bedroom, eight by ten, with a partition between it and the business area, which took up the rest. This room contained an oil-burning range, a sink, and a work table. Pots, pans, ladles, cleavers hung from spikes pounded into the ceiling. This was where Red worked his magic.

Then there was the customers' area. This consisted of a plank counter eight feet long, with four stools. Each stool was of the same design—an upright two-by-four, three feet high, with a one-foot two-by-four spiked crossways on top to sit on. It looked like Red was expecting parrots instead of people. Big parrots. The groceries were kept under the counter, except the butter, which was kept in a bucket lowered into the well alongside the creosote pile.

To get to the door of the diner you had to walk along a boomstick connecting it to the government float. There was no handrail. The type of caulk-boot clientele Red catered to didn't need one.

The deputy minister of transport did need one, although, with much shoving and tugging and reeling and whooping, we finally got him aboard and positioned him gingerly upon one of the parrot perches. Red turned from the range dressed in his usual uniform, white pants and more or less white singlet, showing bare arms and chest bristling with curly red hair, topped by a shiny bald head. He spread his hands on the counter and leaned right over, face-to-face with the deputy minister, and bellowed, "Alright, you bastards, whaddya want? Ya can have ennything ya want as long as it's bacon an' eggs. Ya wannem over or straight up? Cuz you're gonna get 'em straight up!"

With that he wheeled around to his stove, grabbed a handful of eggs, splashed them into a sizzling black frying pan by crunching them in his fists two at a time, slapped a bunch of bacon onto a smoking griddle, slid everything onto plates, stripped the fat off the griddle, flapped down half a dozen slices of white bread, flipped them once and presto! Bacon and eggs and toast à la Red. Fortunately, the deputy minister of transport was very hungry.

TWO

The Governor

I SPENT THE FIRST FORTY YEARS of my life with people like Red
Mahone, who populated a community of loggers, fishermen, stump
farmers, beachcombers, hermits, renegades, remittance men,
confidence men, Greek scholars, stagecoach robbers, and outright
lunatics strung out along the beach from Vancouver to Queen
Charlotte Sound. It was a community that surely must have been
unique in the history of man. I didn't realize it then of course, and
later wondered what was wrong with my head that I didn't take more
time to appreciate it while it lasted.

I know what it was with me. I was too busy trying to make a buck.
The things I paid most attention to were things that came my way in
the course of my various enterprises—first logging, then fixing and
making radios, then flying planes. Sometimes I cast my mind back
to those times, trying to remember a face or a place, but I can't. I
had business on my mind and was too busy to stop and investigate
everything along the way. What I thought so important then seems
inexcusably trivial now. Fixing someone's radio so they wouldn't
miss that week's *Jack Benny Show*, probably. Trying to get on my
way to the next bay, where someone else had a couple hours' work at
75 cents an hour waiting for me. I was one of the coast's original
workaholics.

If you go right back to the beginning I suppose you could blame the
Governor. The Governor was not another potentate of Ottawa
officialdom, but the head of the Spilsbury clan back in Derbyshire.

You see, we were English. The Spilsbury seat was in Findern, Derbyshire, where the head of the family, this would be my grandfather, was the Church of England clergyman and the rest of them were—gentlemen. They were what you call landed gentry, meaning they just lay about doing nothing, living off a community of tenant farmers. It was considered below them to even associate with anybody who did any work. So you had my grandfather, whom everyone called the Governor, four daughters, and five sons: Frank, George, Ben, my father Ashton, and the youngest, Humphrey.

God knows what ever possessed him, but Frank, the Governor's eldest son, decided to leave the old family place. He never did anything else in his entire life that I know of, but he must have got the bug to be an adventurer or something, and in 1878 he came out to North America. He took the Union Pacific train out to San Francisco and then came up the coast by paddlewheel steamer, arriving in New Westminster. This was eight years before the CPR came through. Whether he had heard from somebody who'd been out here or whether he just took it upon himself to stop drinking long enough to get here I don't know. Dad could have told me. I never thought to ask him.

Being an Englishman of the leisure class, hunting was in his blood. Not riding to hounds. I don't think they made horses big enough for Uncle Frank to ride. He was enormously fat. He hired an Indian and canoe to paddle him up the Fraser River. Today the area he travelled through has a population density exceeding that of Holland, but there was hardly a sign of development then except around Fort Langley, where a few Scottish farmers had come out with the Hudson's Bay Company and married Indians. Something made him stop and camp where Whonnock now is. I can remember seeing the split-cedar one-room shack right on the riverbank where he had lived. Being the eldest son, he could get money whenever he needed from his father and he had Indians waiting on him hand and foot like coolies. He took unto himself an Indian mistress and he stayed there a number of years. And hunted. Years later my uncle August Baker, who was half Indian, would describe in his very quiet way how Uncle Frank would hunt. He would go up and sit on a rock with a double-barrelled shotgun. Not a rifle. He loved hounds and always had a mess of them around. He'd send Indians back in the hills to round up the deer and drive them down the trail and he'd bang them off as they ran by.

After three or four years he got tired of this and wanted to go back home. But in the meantime he had put a piece of his inheritance into two quarter-sections of land, and he couldn't just walk away from it, so he undertook to entice one of his younger brothers to come out. The one he picked on was my uncle Ben, who was a bit older than Dad. Uncle Ben had just finished going to Cambridge. I don't know what he studied there because he never did anything either, although he had a great many silver trophies for sports. Being a gentleman, there was nothing he *could* do, but there was nothing to stop him becoming an adventurer and following his older brother out to Canada. Uncle Frank persuaded him by letter to go to the Governor, draw his money, and use it to buy out the Whonnock estate. The CPR was in by this time so it was easier for him and my fat uncle Frank showed him the ropes, took his money, and went back to England where he took up residence with a wench in a disused windmill-tower and lived to a ripe old age.

It wasn't very long before Uncle Ben, who didn't like physical work and was fond of his comforts, decided he wanted to go do something else with his money than camp on the riverbank at Whonnock. There was still one brother left who hadn't been wised up and soon my dad, Ashton Wilmot Spilsbury, was getting letters telling him about the marvellous country here—the Indians, the deer, the adventure, etc. Dad was down at Cambridge taking medicine and Uncle Ben couldn't wait for him to be through.

The thought of all this sport and adventure appealed immensely to Dad, but he had another reason for considering the move to Canada, as he told me many times. My dad was the most unselfish person I've ever known in my life and the thing that was on his mind more than his own prospects was the state of the family fortune. The Governor was getting on, Lloyd George was now on the scene making noises about a thing called the land tax, and the old boy was getting very short of money to keep the whole kaboodle going. There were eleven of them altogether and of course none of them could do anything except sit around showing off their breeding and eating up the family fortune. My dad knew it was costing more to keep him at Cambridge than the old boy could really afford, so after only two years of medicine he got his inheritance money and came out and bought out Uncle Ben.

After showing Dad the ropes Uncle Ben took his money and teamed up with R.V. Winch. R.V. Winch was in the insurance business, the real estate business, the land business, he operated a string of canneries up and down the coast—for half a century he was one of the biggest movers and shakers in this part of the world. The Winch Building at Granville and Hastings, famous as the site of the Vancouver Post Office riot in the thirties, is still a Vancouver landmark. Even in those days old Winch's worth was in the millions, although he had no schooling and was said to scrawl an X for his signature. Then came the great depression—not the one you and I think of, this was in the 1890s—and Winch went absolutely broke. He lost all his money and lost all my uncle's with his. So Uncle Ben followed his elder brother Frank back to Derbyshire.

My dad invested all the money he had left in clearing. His total holding was 360 acres and he cleared 45. The spruce trees and the cedar trees down on that bottom land beside the Fraser left stumps that were twenty feet across the roots and they'd blow them with black powder, then twist them out of the ground with a stump-puller and a team of oxen. It was terrible work. A lot of money went into drainage—he had to put in a tremendous system of ditches. He hired gangs of Chinese and East Indians, but by the time he'd spent all his money he had a marvellous farm. There were over forty acres cleared and over twenty acres under intense cultivation. Dad was very scientific and was going by all the latest books in making it the most ideal farm in this part of the world.

Dad met my mother, Alice Maud Blizard, when she came out from London to spend the summer with her elder brother in Fort Langley in 1893. They were married in Vancouver in 1898. He was 26 and she was 27.

And where did they go for a honeymoon? To Lund. A tiny little outpost on the coast a hundred miles north of Vancouver. Dad and Uncle Ben had a Swedish housekeeper early on, Maria Johannson, who'd left to get married, and the man she married was Charlie Thulin. This was the Thulin who, with his brother Fred, had founded Lund in 1889 as a place for steam tugs to get water and cordwood on their way up the coast. Lund is named after a university town in Sweden, but not, as has often been said, the Thulin brothers' home town. They picked the name Lund simply because they thought it would be easy for Canadians to say. They were practical, resourceful men and soon added a small store, hotel, and bar facilities to their new Lund—all the amenities needed to satisfy the towboater's body and soul. This housekeeper had written Dad about what wonderful country it was, so he and Mother decided, what the heck, let's go have a look.

They went up on the *Tees,* a steamer put into coastal service by the

Canadian Pacific Navigation Company two years previous. Captain John Irving intended her for use on the west coast of Vancouver Island, but in 1898 he had her on the Alaska run, overloaded with horses, dogs and gold-crazed miners on their way to the Klondike. The shores of the south coast were for the most part a blanket of green, unmarked by settlement. There were a few buildings along the beach at Gibsons Landing, which had been established in 1886. At Sechelt, Bishop Durieu's model Indian village, with its pretentious twin-spired church, was in full flower. There was nothing at Powell River and wouldn't be for another ten years, but across Malaspina Strait, Vananda was a booming mining centre with three hotels, a hospital, saloons, and an opera house.

After paying their respects in Lund, Mother and Dad rented a double-ended rowboat built by Karl Anderson, rowed over to Savary Island and spent a week there, taking in the peaceful Gulf Islands atmosphere and admiring the beautiful white sand beach. In 1898 the only building there was an empty log cabin which had been built much earlier by an enterprising individual named Jack Green, who had operated a store at the east end of the island, now called Green's Point. Now the island was once again abandoned, Green having been murdered in 1893, and Louis Anderson having not yet arrived. Mother and Dad stayed in Green's cabin, which still had canned goods and cooking pots strewn around. The next night they spent under canvas at the other end, Indian Point, where they caught some salmon which they "planked" Indian style. Mother also brought down a small deer, remarking how tiny it was compared to the mainland variety. After enjoying Savary they casually rowed across Georgia Strait and spent a week on Vancouver Island.

At Oyster River, halfway between the present-day towns of Campbell River and Comox, they met an ancient squaw wandering through the bush with a very severe stomachache. Dad gave her a handful of laxative pills and they never saw her again. Dad said he was just as glad. She told them she was the sole survivor of an attack on her tribe by the Kitimat Indians, who snuck up at night and murdered them all in their sleep. She awoke during the attack, slipped away between the wall planks and spent the night hiding up to her chin in water. They were never too sure what to make of the old woman's tale, but I have since heard of this raid from other sources. She did survive Dad's doctoring apparently, as she was glimpsed living alone in the bushes for many years after.

Apart from her my parents saw not a living soul, and gloried in

The Longlands, home of generations of Spilsburys in Findern, Derbyshire, about the time I was born there in 1905. That's the Governor in the doorway.

The Spilsbury clan circa 1903 doing what they were good at. Sitting: Aunt Kate, Uncles Frank and George, my grandmother and the Governor, and my aunt Bess. Standing: my uncles Humphrey, Ben and August. Aunt Bella probably took the picture, and my dad was in Whonnock blowing stumps.

My dad's 45-acre farm at Whonnock circa 1906. You can see my fat uncle Frank's split-cedar shack on the river bank.

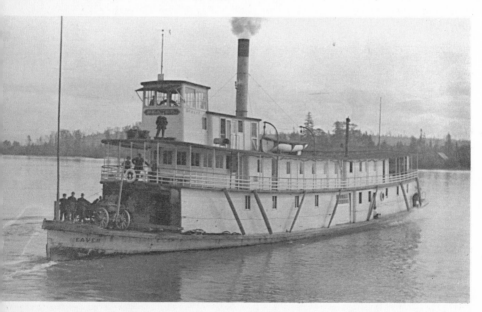

At first, the only link with the outside world was the CPR paddlewheeler Beaver, *seen leaving the farm here.*

the primitive solitude of the river. In 1904, six years later, Charlie Thulin would come over and found Campbell River as a refueling depot modeled on Lund, and their ex-housekeeper would become the *grande dame* of a much larger settlement, but at this time no one lived for many miles around. Dad was never high on killing for sport, fish or fowl, but Mother was a great outdoors person and a real huntswoman. While they were there she shot several deer, a bear, and a cougar — and not with the help of Indian beaters herding them down the trail. The skins of the cougar and the bear cluttered up our home for many years, draped over the backs of chairs. As a kid I hated their smell and was terrified of being left alone in the room with them.

Mother and Dad spent a better part of a month rowing around the northern Gulf altogether. It must have made a profound impression on them because they resolved that some day they would sell the farm, have a motor launch built, and return to cruise up and down the coast in style. In the meantime, they worked the farm together with great dedication. They built up forty head of Jersey cattle. They would twice a day take their milk cans in the horse and buggy to meet the milk train on its way to New Westminster. They made butter, they had a large orchard, and they specialized in the manufacture of Devonshire Cream. They hoped to make it popular in New Westminster, but it never really caught on. They were one of the first to put in a modern cream separator, where you wind the big handle and the little spinner in the middle starts speeding up and speeding up and speeding up and you have to run it at a certain speed while someone else pours the milk into the hopper. The heavy milk goes to the outside while the cream, being lighter, runs out a spout from the centre and the milk runs out the outer spout. You could regulate it accurately to any percentage of butterfat that you needed. A very scientific device, for those days. Dad was the first in the area to put in silos. I'm old enough to remember them being built. They're everywhere now, but when he did it they were strictly experimental.

In 1905, when it became apparent I was on the way, the family in England decided no Spilsbury could be born in the colonies and commanded Mother and Dad to come home. With these people, when it came to a question of doing right by king and country, continents couldn't stand in their way. Dad was a bit of a free thinker in some ways, but he was very loyal to the old flag. When Queen Victoria died, he was on the municipal council, and the

records show that he made a motion that the municipality of Maple
Ridge, British Columbia, send their official condolences to the royal
family. Now he unhesitatingly placed his precious herd in the care of
a neighbour, loaded Mother and himself on the CPR to Montreal,
and sailed across the Atlantic. They duly presented themselves at the
big old family home in Derbyshire in time for me to appear October
8, 1905, in the same upstairs bedroom where Dad had been born.

The family's sense of propriety was satisfied, but at my parents'
expense. Dad had planned to come right back to Canada, but shortly
after he arrived in England he came down with a serious bout of
rheumatic fever. It was touch and go for him, and he had to be taken
into the city of Derby to undergo special care that lasted more than a
year. Mother was meanwhile pinned down in Findern with this crew
of genteel snobs whose code of conduct was terribly refined when
dealing with others of their class, but quite different with someone
they perceived to be below them.

My mother came from a merchant family in London. She
somehow contrived to be born in Tipperary in 1871 and liked to give
the impression she was Irish, but she wasn't a bit. Her family was
descended from the Huguenots. Her father was in the garment trade.
Quite well off, but that didn't make up for the fact that he actually
earned his own living. The Findern bunch owed their lofty position
to an ancestor who made a fortune digging ditches, but that was long
enough in the past they could ignore the fact. They considered
Mother a caste below them and they treated her accordingly.
Apparently my uncles had the reputation of being woman-chasers
who'd get the chambermaids into trouble and then hush it all up.
Rightly or wrongly Mother got the impression she was being placed
in the same category and she resented it bitterly.

My aunt Edie, Uncle Ben's Scottish wife, was a wonderful person.
She was the quiet, observing type, aware of everything that was
going on around her while managing to avoid getting involved in the
hassle. She maintained that my mother's experience during the stay
with my father's family in Findern had such a traumatic effect it
changed her whole life—and Dad's life too. That sort of treatment
would have been hard for anyone to take, but for my mother it was
much, much worse. Whether it was her French blood or what, she
reacted violently, being very fiery and proud in her own right. She
came away from this spell in England just searing with rage. It
stayed burning inside her, and the form it apparently took was to
turn her against everything that people like my father and his family

Whonnock farmhouse built by Dad and Uncle Ben in 1888.

Mother was considered a bit of a looker by 1890s standards.

My dad as a Cambridge medical student.

stood for. She cut her hair short like a man's and took to wearing men's trousers. To show her absolute disdain of everything proper and British, she became an ardent suffragette, she adopted the cause of the anti-British terrorists in Ireland, and generally became a very difficult person to live with. She led my poor dad a hell of a life, really, and according to my aunt Edie, it all started with this period they spent in England at the time of my nativity. And mine was the last nativity that took place in our family.

By the time poor Dad got back to Whonnock he must have wondered what hit him. A year and a half earlier he had been on top of the world with an exemplary farm and an energetic, supportive new wife. Now he had a farm that needed new fencing, new ditching, and new stock, an order from his doctors that forbade him to attempt any exertion more strenuous than a slow walk, a wife who was severely disenchanted, and a squalling new dependent in the form of myself.

That's how I came to be born in England, an Englishman. I've been trying to live it down ever since. But I started out to explain why it was the Governor's fault I spent my life trying to make a buck.

One of the Spilsbury ancestors was a big-time contractor and he constructed a lot of the canals throughout England in the period before railways, when the whole country was being opened up by a network of waterways. He made a lot of money. A very capable man apparently, by the name of James Ward. I wish I knew more about him.

When I was being born over in Findern with all the clan elders in attendance, it fell to the patriarch to select a name, and what with these financial difficulties he was having, my dear old grandfather leaned forward on his cane with his flat hat and everything and rumbled, "Name him after the only member of this family who ever made any money. It's high time somebody else made some money!"

So I was named Ashton James Ward after this ancestor who founded the Spilsbury fortune in transportation back at the start of the Industrial Revolution, and in a family like that, there was much in a name. I was aware of this from the time I was aware of anything. People were always setting it in front of me. Even my mother, oddly enough. She would say, "Now, your grandfather mapped this out for you, and you mustn't forget." Everybody went along with it, the whole tribe of them. Every time I did something, like drawing a vaguely recognizable picture of a steam engine or

The Governor around the time he named me.

I got farming out of my blood early.

graduating from grade four, my aunts would write and say how proud they were to see me getting on with making a success of myself. They kept harping on this. I grew up feeling it was up to me to get out there in the world and move mountains like James Ward.

THREE

Savary Island

DAD HADN'T THE LUXURY of being able to follow his doctors' orders and immediately set about working sixteen hours a day rebuilding the farm. I don't know whether his close brush with death had got him in the way of thinking he should get on with his real ambition, or if perhaps he thought it would make Mother happier, but he also started work on the boat in which he hoped to cruise up the BC coast to Lund and Desolation Sound and everywhere they'd been on their honeymoon. He sold the farm to a large family of the area by the name of Watson and got another neighbour who was a boatbuilder, Olas Lee, to make a scale model of their dream boat. Thirty-six feet long, compromise stern, with the usual raised pilothouse and low aftercabin — I could still draw you the details. It was to be powered by a two-cylinder, heavy-duty Corliss gas engine. Make-and-break ignition. The very latest thing.

The first thing that went wrong — Olas got into trouble. I was too young to be told what it was, which makes me think it involved a girl. The rest of his family, being upright sorts, got him to hell out of there. Dad had made a $500 down payment on the engine. I remember going into a little shop in Vancouver on the water side of Georgia Street near the park and seeing these great, green-painted, heavy-duty engines for sale. "Well, when's your boat going to be ready? We're ready to install it." Dad took his plans to another boatyard, but the people who bought the farm stopped making payments and he ran out of money.

By now it was 1914, and war was declared. In World War I,

anybody serving overseas didn't have to meet their debts. This Watson, who had the farm, was much too old to be in the war but he had something like eight boys, two of whom went overseas, and that was enough to allow the whole family to claim protection under the wartime moratorium on debts. It was 1922 before Dad finally was able to evict them for nonpayment. By that time the ditches were filled again, the fences were down again, the cows were all gone again, and the taxes were in arrears. Dad had borrowed the money to build the silos from the Westminster Trust company, and one day he simply got a letter saying they'd sold the whole farm for what he owed on the silos.

The last time I saw Dad's boat the keel was laid and the ribs were up. He lost about two thousand dollars on it and it cleaned him out. That was the end of his dream. Dad was the only Spilsbury of that generation who really did any work and he was the only one who died without money.

During the ten years when the farm was hung up Dad became desperate for a way to earn a few dollars to keep a roof over our heads. We moved into a tent. An elderly gentleman named Colonel Herchmer, who had come west to set up the Northwest Mounted Police and shared with Dad an affection for Savary Island, took an interest in our predicament. The Herchmers had built a cabin on Savary where they spent their summers and they invited us to go up and stay in it for the winter. There were enough people staying over the winter that year, 1914, that for the first time they were able to have a school, and as I was already three years past school entrance this was very appealing to my parents. Dad could do some work building the fences for Colonel Herchmer and earn a few dollars, so we made a temporary move to Savary Island—and stayed there for thirty years.

Savary Island has a history unlike any of the other larger Gulf Islands. It attracted the attention of settlers early on because of its lovely sweeping sand beach and its low-lying land, but it never really took as a year-round community. It's not suitable for farming. There's no grazing land to speak of. It's sandy soil. You're limited in water to what you pump out of a well at mean sea level. It's an overgrown sandbar is all it is. Even the loggers who logged it over and over all went broke.

The first actual settler was undoubtedly Jack Green, an Englishman from Hull who probably began living on the island

Lund in the early days.

The log cabin store where Jack Green was murdered.

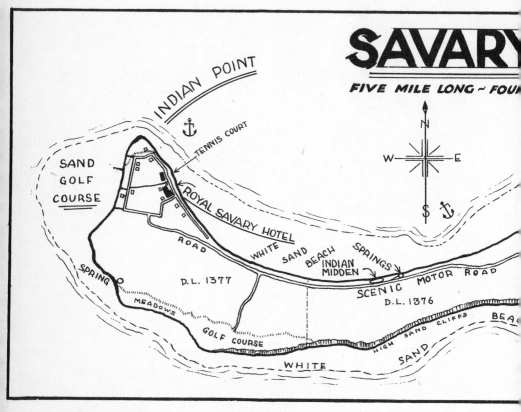

I drew this map of Savary in the thirties, at a time when there was far more development than when we arrived in 1914.

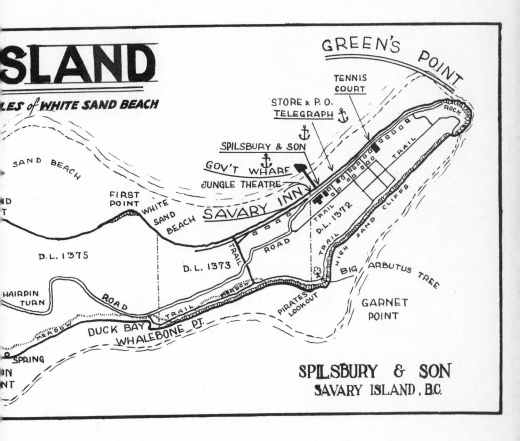

SLAND

LES of **WHITE SAND BEACH**

GREEN'S POINT

TENNIS COURT

STORE & P.O.
TELEGRAPH

SPILSBURY & SON

GOV'T WHARF

JUNGLE THEATRE

SAVARY INN

FIRST POINT

WHITE SAND BEACH

SAND BEACH

D.L. 1375

D.L. 1373

D.L. 1372

HAIRPIN TURN

ROAD

ROAD

TRAIL

TRAIL

TRAIL

TRAIL

TRAIL

ROCK

HIGH SAND CLIFFS

BIG ARBUTUS TREE

PIRATES LOOKOUT

GARNET POINT

DUCK BAY

WHALEBONE PT.

SPRING

SPILSBURY & SON
SAVARY ISLAND, B.C.

Savary Island looking over Indian Point toward Green's Point.

Savary's famous beach looking toward Green's Point from the wharf.

Our tent just after we put it up at Blair Road in 1914.

Me aged about ten studying in the driftwood living room.

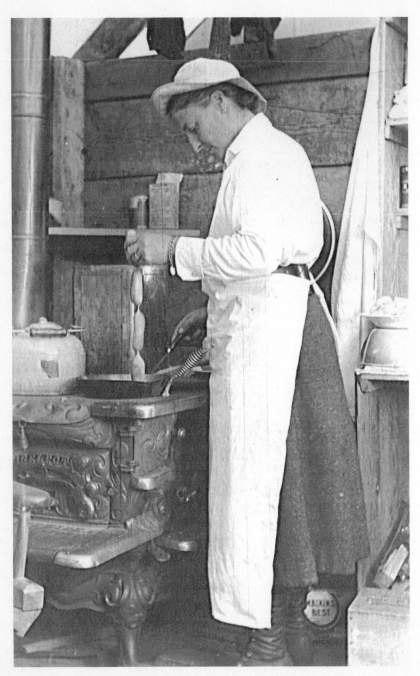

Mother fries up a mess of sausages in the canvas kitchen.

several years before he pre-empted 160 acres in 1888. Green ostensibly ran a store. In his day much of the travelling on the coast was done by means of rowboats. Handloggers thought nothing of rowing back and forth to "town," whether it be Nanaimo, Victoria or New Westminster, for their annual supplies or to have a real binge after selling their logs. Green's store was a handy place for them to stop en route and buy the odd thing, including a bottle. The sloping beach was a handy place to drag a canoe or rowboat out of the water for the night. He would issue credit on occasion when a thirsty man was short of cash, and Green would permit him to work off his debt on the "estate." Green kept increasing his holding until by 1893 he owned four-fifths of the island—958 acres. He cleared the whole east end of Savary from shore to cliff for about half a mile west to a clearly defined line running across the island, along which he built an old-style snake fence. This permitted him to graze a few cattle.

When I was a kid this whole "farm" area was thickly covered with second-growth fir trees up to six inches in diameter, and counting about 25 or 30 annual growth rings. Parts of the snake fence were still intact but hidden by the new growth. Numerous fir stumps with springboard notches remained from Jack's original land clearing and logging efforts. According to old Louis Anderson, who was the next permanent settler, Green at one time used a team of oxen to put some of this in the water, but it is unlikely he got very much for it. Savary Island fir trees have never been good grade, and in those days would not fetch much of a market. Nowadays it is difficult to see just where Green's fence line used to be because the adjoining property has since had most of the original trees cut down, but in my day it was very noticeable and could be seen from the mainland. Everything east of the fence line was bright green new growth, while to the west it was the darker green of mature trees.

The story of Jack Green's log cabin store, his murder, and his lost treasure are officially recorded in the annals of the BC Provincial Police in Victoria and have been written up many times. A man by the name of Hugh Lynn, originally from Lynn Creek in North Vancouver, was apprehended in Washington, extradited, tried, found guilty, and hanged in New Westminster. The old whitewashed log cabin which served as Green's store stood empty for many years. As kids we would play in it. People camped in it for short periods of time. For half a century, the whitewashed gable of the old cabin served as a landmark. The east end of the island was always known as Green's Point, although some civil servant has since stroked that fascinating history from the map and recreated it as Mace Point, honouring a much later settler. There was an old well behind the cabin, long since filled in. The remnants of the old cabin were considered a hazard for children and it was torn down and burned by residents in the late thirties.

The story of Green's buried treasure persisted for years, and I can remember seeing the whole ground dug up around the cabin like a pack of gophers had been at it, but as far as I know no one ever found anything. Playing on the gravel beach right in front of the cabin, though, I once found an English copper penny I still have.

After Green was murdered, no one lived on Savary for several years until Louis Anderson built his first house, which still stands, about four hundred yards west of the present wharf. Louis was originally a logger from Sweden, but at the time he moved to Savary he was a timber cruiser for Hastings Sawmills in Vancouver. I don't know why he chose Savary, or whether he actually owned the land he built the house on, but after the coming of the Savary Island Syndicate in 1910–11 and the much-publicized real estate development he sold the old place to G.J. Ashworth, later known as Captain Ashworth, the founder of the Royal Savary Hotel and father of the more recent owner, Bill Ashworth. Anyway, at this time Louis Anderson acquired four lots and built another house which looked just like the first one, but he burned it down around 1918. Then he built another, later sold to Robert Cromie of the Vancouver *Sun*.

In the period from, say, 1912 to 1920, there would be an average of five or six families at the most who could be considered permanent residents of Savary Island, plus the odd drifter. Maximum population in the winter time was about thirty to forty. Maximum school attendance during 1914 to 1918 when I attended was fourteen pupils; seven or eight would be a closer average.

We had to get out of the Herchmer's cabin, and Dad didn't have enough money to go anywhere else, so we set up our tent and squatted on an unused road right-of-way. In 1910–11 the whole island had been surveyed into small lots by the Savary Island Syndicate, a get-rich-quick scheme inspired by an engineer in Vancouver called Harry Keefer. Some draftsman sitting in a warm building in the city had drawn road right-of-ways straight up hills and down cliffs—they could never be built in reality, but on the maps they were legally set aside. We put up our tent on an imaginary thoroughfare called Blair Road, which came plunging down a precipice just west of the wharf. We squatted there ten years. It was 1924 before we got back in a house.

Together with the time in Whonnock, I've spent twelve years of my life living in tents. Mind you, they were damn good tents. We had two set up. They were ten by twelve with a space between and the fly covering the whole length. The platform consisted entirely of planks picked up on the beach. Rounded, and weatherworn—there were four-by-tens and ten-by-twelves—anything we could find we'd tow home, drag up the beach, dig into the clamshell, and level off. We had a floor then that would last almost forever and the tent frame was built on top of that. We fixed it up very nicely. After a while it got pretty breezy so we got driftwood and boarded up the back and then part of the front and had a door and put windows in. Then the fly rotted, and we replaced that with split shakes. It ended up the tent was more like wallpaper; we built all around it.

It was really very cozy. It couldn't have suited my mother better. She loved to think she was roughing it. Dad had his books and I remember him sitting there evenings reading in his black Cambridge blazer with the Clare College crest, just as happy as if it were a castle. He always kept a brew of ale going, not that he was a heavy drinker, but he liked to entertain and he'd have all the neighbours in to talk. He'd spend long hours debating theology with the Anglican church minister, Alan Greene. Greene wasn't a deep thinker and accepted the Word the way the church required him to. His was not to question why. Dad was a very doubting, non-churchgoing Christian who also read the latest scientific literature and tried to reconcile this with the Bible. They would go on and on together.

Greene became quite a landmark on the coast, mostly because he lasted so damn long. He had married and buried generations of oldtimers by the time he was done, sometime in the 1970s. I vividly remember the first time he showed up at Savary. He came out as a student with John Antle.

John Antle was the founder of the Columbia Coast Mission, an Anglican Church organization serving the mid-coast area. He was a Newfoundlander and a good seaman with a grasp of practical matters, and he did good work on the coast setting up his loggers' hospitals at Alert Bay, Rock Bay, Vananda and Pender Harbour and on board their ship, the *Columbia,* but he was never well liked. He lived on Savary for a year, staying in the Ashworths' cottage. He had a daughter our age, Marian Antle, and I think she went to school there one year, then they moved across to Whaletown on

Cortes Island. Mrs. Herchmer was very churchy and kept Antle coming back to give services in her home.

Before this particular church service they'd had a telegram from Whaletown that Antle was sending Mr. Greene in his place. The mission had a small boat twelve feet long, carvel built, painted grey, with a little two-cycle Easthope in it, pupupup. He came down through Baker's Pass and headed into the beach at Savary, where we kids were all out looking for him. In fine weather the westerly breeze always gets up in the evening and on this day it hadn't been quite calm to begin with. By the time Greene arrived the seas were running about two feet high. There was no float for small boats so he headed straight in toward us and shut the engine off. He came in on a breaker, the bow stuck in the sand, and the next wave took the boat and rolled it right over on top of him with all the prayer books and the foot-operated reed-organ and the whole works. We dragged him ashore, dried him out, dried out his gear, and that was how we met the Reverend Alan Greene. He knew nothing about boats, and to the day he died he was still not a seaman. The church is pleased to think that he became a great navigator, but there is hardly a rock on the BC coast he wasn't up on at one time or another. However, he had a great sense of humour and would tell wonderful stories on himself. He was very sincere, had a great faith in the Lord, and came to be very much loved by all of us.

He was way out of his depth with Dad. Dad knew his Bible from cover to cover, in the Latin and Greek versions to boot. Dad would go back to the original Bible and cite earlier translations where it had totally different meanings. Greene would be completely stumped. I don't think he allowed himself to think deeply, but he didn't mind getting into all sorts of discussions as long as the ale supply held out.

Savary at first didn't offer a great deal of scope for working out my destiny as a captain of industry, but for the first few years I was busy enough catching up on my formal education. I started school in September, 1914, a month before my ninth birthday. Since I had taken instruction at home under my Dad's tuition I could handle the three Rs. I was accepted into Lower Second Reader and slugged my way through to Upper Fourth and High-School Entrance by the fall of 1918. As far as I can tell now, during those four years of "formal education" I learned absolutely nothing I didn't already know. I didn't even enjoy the experience. On the contrary, they were undoubtedly the most miserable four years of my life.

Everybody turned out to meet the weekly steamer on "boat day."

High-fashion swimwear in 1915. Me in the stripes, Laurencia Herchmer in the bloomers, Mother with the dog, Miss Burpee next to her.

In 1915 the talk was all World War I, and so was our play. I'm second, Jimmy Anderson fourth. Our leader is Frank Turnbull.

You can see where Dad got the idea I'd be an engineer.

The teacher, Miss Frances Stevens, spinster, aged 28, recently from Edinburgh, would summon us in the morning by leaning out the door and brandishing a hand-held bell. I understand this artifact is now on display in the Powell River museum. The one I remember was no ordinary bell like you see country schoolteachers ringing in pictures. This was a double gong sort of affair. It consisted of two saucer-shaped brass bells with their concave sides facing, bolted together at the centre. The clapper was inside, and the hardwood handle projected radially from this clamshell arrangement. They were intended to "resonate" and sound melodious and beautiful. But to me it was the bells of doom and nothing less. My recollection is that Harry Keefer, by then a full-time island resident serving as both postmaster and storekeeper, loaned it originally to the school during the school year, but that it belonged to the Savary Inn, which establishment he was also running in those days. Keefer was also good enough to provide us with a teacher's strap. I have reason to remember this piece of equipment in great detail. It was leather, originally black but worn to a dull brown in spots. It was three-quarters of an inch wide and about twenty inches long, with a buckle on one end. It came originally from around the neck of Harry Keefer's white horse. Miss Stevens showed her gratitude and appreciation by hanging it on a nail behind her desk for all to see. From time to time she would glance at it with what was unmistakably keen anticipation.

We didn't have very long to wait. I was the first one to get it. I have forgotten the provocation now, even if I understood it at the time, but I clearly remember the incident. She gave me the usual tongue-lashing, then hauled me up in front of her desk in full view of the whole class and told me to hold my hand out toward her with the palm up. She was very shortsighted and used very strong pince-nez glasses. She took two swipes at me and missed both times. The class started to titter. She took her glasses off and laid them on the desk in front of her, took a fresh grip on the strap, and this time she nailed me. It was more painful than I had any reason to expect. Now that she had the range, she took another swing with all she had behind it. I involuntarily drew my hand away, and the strap came down on the desk top with a crash, scattering various pens and pencils and spilling Carter's blue ink all over the place, but when she found that she had broken her glasses, she *really* blew her stack!

Poorly concealed sniggers from the rest of the class didn't help my situation at all. She made me stay in after school for an hour, making hundreds of copies of some appropriate sentence while she wrote a long and damaging report for me to take home to my parents. Whatever my alibis were, they apparently didn't carry much weight, because Dad took me out behind the woodshed and administered a real tanning with a three-foot switch, and not on my hands either. From that time onward, strapping became more or less routine in my case.

It was about this time that I became disenchanted with a scholarly career. Maybe I'm being unfair. Maybe I didn't realize the problems that faced a teacher in that school. Imagine fourteen pupils spread over seven classes, all in one small room. I graduated from the Savary Island School in July, 1919. I didn't know what the hell to do because to go any further I'd have to move off Savary. The closest high school was in Powell River. My parents weren't about to go there, so that was the end of my formal education.

I have a certificate saying I can attend any high school in the province of British Columbia.

You might think my father would have been disappointed, being a Cambridge man and all. His great hope early on had been that I would become a civil engineer, end up going to South America and building bridges. He was very keen on the engineering side, and I thought that was what I wanted. Of course, without a high school education, let alone a university education, it was out the window. But if he was disappointed, he never let on. Dad didn't put too much faith in the surface of things. He felt as long as you had a good mind and you knew how to think, you'd get there. Even before I went to school he had taught me arithmetic, he taught me to read, to write, he taught me algebra, he taught me a little trigonometry and a little Greek.

Dad worshipped the word *logic*. He loved mathematics. He was a great admirer of the scientists of the time — Rutherford, Planck, Einstein. He admitted Einstein's mathematics were beyond him but he'd talk about the theory of relativity for ages. He was terribly thrilled with all of that.

The use of reason, how to analyse problems logically, was the lesson Dad always came back to, because he felt if I had that I was as well equipped to face the world as any university could have made me. Law and order and reason, that was his credo. Never get excited. Mother thought he was so bloody dull, that's what bugged her. She liked to get all hot and bothered and excited and jump around — real French as I see it now. I guess I had a little bit of both.

FOUR

Wild Animals and Wilder People

PEOPLE ON SAVARY GOT BY, as the saying goes, by taking in each others' washing. Apart from building and repairing the summer cabins there was well-digging, working on the road and cutting wood at $3.50 a rick split and piled—we'd buck it on the beach with a hand saw and wheel it to the customer. There weren't many ways of bringing in the groceries. This resulted in the people of the island developing a special relationship with the area's wild game.

The various methods employed to procure game and their relative success rate became matters of great personal pride—either proud of the way they went about it, or too proud to admit to the practice followed. Take pit-lamping, for instance. This was a surefire method of getting deer by shining a light in their eyes at night. It was illegal for the good reason that it was difficult to tell whether the eyes reflecting in your light were attached to a deer or your neighbour's cow, or perhaps to your neighbour himself. But it was very effective and some, like old Louis Anderson, wouldn't waste their time doing it any other way. Besides, Louis was an old-time prospector and legitimately owned a miner's lamp, the kind that used acetylene and water and was normally worn on a man's hat. This was ideal for pit-lamping since it left both hands free to operate the rifle.

People devoted to this sport scoff at the suggestion they could ever mistake the way a deer's eyes shine in the dark from, say, the way a cow's shine, but I'm not too sure. One day Bill Ashworth and I nailed a couple of beer-bottle caps about four inches apart to a

21

stump at the top of the logging road, which was a favourite deer hunting spot. Next day there were bullet holes all around them. I never did understand why any hunter would empty an entire magazine and not get wise to the trick. Of course no one ever owned up to it, so it remains a mystery to this day. Since pit-lamping was (and is) so decidedly illegal, great secrecy was always observed. Those who did it never talked about it. Those who didn't expressed their disapproval vociferously. My mother was one of these. On the other hand, when someone offered her a haunch of venison she never enquired as to the time of day it was shot.

My mother was well known for her success in the field. She regularly shot her limit every year, but she had her own methods. She hated rifles of any kind—even my little single-shot .22 caliber Winchester. According to her, all high-power sporting rifles should be banned from the earth. She used a double-barrelled 12-gauge shotgun for everything: #7 chilled shot for duck, #6 for geese, and SSG for deer. SSG is larger than buckshot; the pellets are about a quarter-inch in diameter as I recall. They sure make a mess of an animal at close range, and she only worked at close range. Her method was to walk over the ground carefully, identify the trails, and then post herself quietly behind a tree and wait patiently until the right deer came along, and when he did, believe me he was dead.

My first and only experience of shooting a deer was a messy business, and I was left with no desire to repeat it, but first I should go back to the circumstances that led up to it.

On a very stormy winter afternoon in 1922, my dad and I came in from working in the driving rain, and were preparing our supper in our old tent/cabin at the foot of Blair Road. Mother was away and there were very few people on Savary at the time. There was a knock on the door and in stumbled an elderly man, very wet and very cold and in the last stages of exhaustion. He could barely speak. He had been shipwrecked the day before on the south side of the island up near the Indian Point end. No one lived at Indian Point, and of course there was no road then. It had taken him over 24 hours to reach our place.

Dad wrapped him in dry blankets and fed him some hot Oxo, and he slept through till morning. After breakfast we put some lunch together, launched our old twelve-foot rowboat, and headed out around Indian Point to the scene of the wreck. At low tide we could trace the whole chain of events, from the big sharp rock at low water level where the bottom was ripped out, leaving a string of nuts and bolts and tools and nails to high tide level, where the only thing left was a small section of the bottom with the engine timbers wedged between two beach logs. The shaft was broken and gone but the little engine seemed to be all there. I was all for attempting to salvage it, but the old man practically flew into a rage and said, "No, no, she nearly kill me! Leave her lay where Jesus flang her!" He meant it.

We did manage to gather up a few of his personal belongings including his gun, which we found buried in the gravel — and what a gun! When I stood it up it was as high as I was. It was a standard issue Swiss Army Rifle. It took a .55-caliber, rim-fire copper cartridge charged with black powder. This was new to me, as the guns I used all took modern cordite-filled cartridges. It had three firing pins and a bolt-action repeater taking ten extra cartridges in the tubular magazine under the barrel. I recognized it because Harry Keefer had one like it which he used occasionally for shooting the bark off dead snags for firewood. The ammunition was cheap. You could buy a box of twenty for fifty cents. The old man asked us to look after it for him till he could come back and get it. I had hopes that he never would, but in fact about a year later Reverend Alan Greene stopped in to pick it up and take it back to the old man on his stump ranch. But in the mean time I had my fun with it.

I made one or two trips back to the wreck site, carefully raked the gravel between the boulders all the way down the beach, and recovered about fifty cartridges, most of which still worked. I oiled and greased the old gun and then tried it out. "Boom!" Perfection! It didn't have too much recoil. The gun was heavy and the black powder cartridges seemed to have greater elasticity than the cordite style. It was like firing an enormous slingshot, except for the incredible cloud of blue smoke. On a still day it required about five minutes for the smoke to clear. You had to walk off to one side to see the target. And range! The rear sight was graduated up to a thousand yards.

There was one other peculiarity. The rusty old barrel had no rifling left in the bore, so the bullet keyholed — it went out end over end and made a big slot in the target instead of a round hole. On a bright day, if you aimed it up in the sky against a white cloud background you could actually see the bullet tumbling on its way, but only when there was a good crosswind to blow the smoke away quickly. Some people wanted to talk me into sawing a couple of feet off the end of the barrel and reshaping the stock to convert it into a sports rifle, but of course I couldn't do this as it still rightly belonged to the old man.

In due course I got the urge to take it out and use it on real game. Dad felt he should come with me just this first time, and it's just as well he did.

There was about six inches of snow on the ground, which made it easy to follow the deer tracks. We went up the back trails and under heavy timber on the south side. It wasn't very long before we jumped a deer, and he ran across our path about fifty yards away. Because of the long barrel and the weight I found it necessary to kneel down and rest my left elbow on my knee and at the same time to pivot to the right to allow for the deer's direction of travel. Dad said I better shoot ahead of the animal by a few feet. It was not easy because the deer was only visible intermittently as he ran behind trees, but I finally coordinated all this and squeezed the trigger. The resulting noise and smoke was impressive, but so was the devastation that greeted us when the fog cleared.

The story was all there on the surface of the nice white snow. It appeared that I had been pretty well on target as far as the deer was concerned, but at the moment of firing, a rotten stump had intervened in the line of fire. From the muzzle of the gun there was a dirty black track of burned powder for about twenty feet to a small hole in the stump on my side. On the far side the keyholing bullet had ripped out about a cubic foot of rotten wood which was scattered over the ground in a fifty-foot arc, on the outer perimeter of which lay the crumpled form of the wretched deer. The bullet had entered his right haunch, proceeded diagonally through the body and out the left shoulder decimating everything in its path and leaving a generous spattering of blood, deer hair, and bone chips for the next thirty feet or so.

One shot, and the place looked like a battlefield! When Dad got through dressing it out we were left with only one haunch and one shoulder to take home and, to make matters worse, it was the toughest venison anybody had ever tried to eat. We ended up putting it through the mincing machine and canning it for dog food. As for the bullet, we never did find out how much further it went, but whatever else it did, it did crossways, that I'm sure of.

Hunting on Savary was not confined to deer. During the winter we had all kinds of wild duck. Mallard, teal, and canvasback were the desirable ones. The others were considered fish-ducks, and shot only

Dad relaxing in his favourite chair wearing his Cambridge blazer.

Wild game beware! Mother in wait with her trusty blunderbuss.

by those lacking in taste—saw-bills, butter-balls, blue-bills, and the common scoter or black duck, not to mention the "kiss-me-asses" and various members of the diver and grebe family.

But then there were the geese, two varieties—the Canada goose and the brant. We still have the Canada goose, especially in Stanley Park and the inner harbour of Vancouver, where they are becoming a real pest, and also at the head of every inlet and river mouth up the coast. They are certainly far from extinction.

But I would like to know what became of the brant. When I was young on Savary, brant outnumbered everything else by a wide margin. While we would often see Canada geese in flocks of twenty or so, brant were out there by the thousands and millions. In the winter they would come in the evening and gather by the tens of thousands out on the reef between First and Second Points, covering an area of several acres—just dense black geese quietly chortling to one another. At low tide, which is always at night during the winter, they would be way out on the beach, in the seaweed and boulders. When anything disturbed them they would take flight *en masse,* with a noise like thunder, and would go cackling away into the distance and eventually come to rest on the reefs off the south side of Hernando or right over to the big reefs off Sutil Point and Marina Island. They would not return to Savary that night. This meant that any hopeful goose hunter had one chance and one chance only on any given night. Many times I have seen the entire flock rise as one bird and literally darken the whole western sky. On any quiet night you could poke your head outdoors and hear the busy chattering of thousands and thousands of brant out on the reefs.

What ever happened to them? During the last fifty years that I have travelled up and down this coast I have never seen a single brant.

Anyway, going back 65 years, the brant was the most sought after and the most discussed bird on the hunter's list, based mainly on their apparently limitless numbers and the extreme difficulty in ever approaching one close enough for a shot. They were far and away the most wary of all game birds on the coast. Whatever fate befell them, I am sure it was not as a result of over-hunting on the coast of BC.

There were several methods employed in hunting any of these birds, depending, I suppose, to some extent on your upbringing. My parents, for instance, would shoot a duck only on the wing. They would never, but never, shoot one on the water or on land. If necessary they would have someone "shoo" the thing up first and then bang away and miss it, or drop it out on the water where they couldn't recover it.

My mother and dad would usually walk away out to the reef off First Point and crouch behind a boulder in the chill rain of a winter's evening, waiting for the evening flight. This is when some ducks were supposed to fly over the sandbar to get to the other side, and if they were within range my parents would blaze away at them. The success rate was minimal.

I guess I came more under the influence of the colonial style. My method was usually to spot a few birds feeding along the shoreline and sneak up on them when they had their heads under water, then freeze and try to look like a boulder when they raised their heads to look around. I would try to get them grouped so that one shot from a 12-gauge would plaster a bunch of them. At best this would work only once on any day because after the first shot all ducks within hearing range would take off for Hernando for the night. The Anderson boys used this method and I learned from them. Their dad was very strict. He would dole out just so many cartridges and they were expected to come back with at least one bird per cartridge. In order to provide for the occasional miss they would always try for two or three per shot when the grouping was good.

The most determined attempt on the brant was made by an individual we knew as Uncle Norman. I think his full name was Norman Thompson. He was related in some way to the Burnets. Ken Burnet, an old-time BC land surveyor, built one of the first cottages down near the Maces, and I believe it is still used by the family. Anyway, Agnes and Lillian Burnet attended school at Savary the first year it ran, and they called him "Uncle Norman." Uncle Norman was something different. He kept very much to himself, spoke very little, and generally minded his own business. He came up to live in Burnet's cottage and look after it one winter when the family was in town. He was a bit old fashioned, we thought. He made most of the clothes he wore out of buckskin which he tanned himself. Buckskin jacket with tassels on the sleeves, buckskin pants and moccasins, and some sort of a leather cap with fur on the outside. He spent most of his time doing this. I can remember he would soak deer skins in some solution until they went partly rotten and then he would spread them over a sort of sawhorse arrangement and scrape the hair off. After this he smoked them and tanned them

Mother hooked a salmon, a cod swallowed the salmon, Dad gaffed the cod, and I got to hold them up. It was 65 pounds of fish altogether.

Me with two of the elusive brant.

with hemlock bark. The smell varied from stage to stage and continued right on as he wore them. It was particularly noticeable when it rained.

Uncle Norman tried all the usual approaches and did no better than anyone else. But Uncle Norman did a lot of thinking about it. He didn't say much but he thought a lot, and finally he hit upon a solution. He lost no time putting the plan into action. He discussed it with no one, but the amount of work required to put the project in motion could not be concealed from prying eyes and everyone talked about it—some were for, most were against. Discussion increased as the work progressed.

In essence, here was the plan. He would build a raft of small cedar logs about four feet wide and six feet long. On this he would erect a dome-shaped hut of bent cedar boughs about four feet high. This structure would be thatched with seaweed and sprinkled generously with an assortment of starfish and barnacles until it resembled a large beach boulder at low tide. He had a flap door like an igloo and numerous peepholes around the sides, just large enough to get a gun barrel through. Between the two centre logs he provided a slot about six inches wide through which he could manipulate a paddle while seated on a small bench. The plan of operation became quite apparent. He would quietly paddle into the middle of the flock and then open up on them from the gun ports before the bewildered geese realized what was happening. He was so confident that he never even took the thing out for a test drive, and there were obviously many glitches that might show up, even from the navigational point of view, let alone the sea- and battle-worthiness of the device. He launched it down the beach on rollers and anchored it off at high tide ready for the countdown. All he needed was the right weather and a reliable goose forecast and he would be away.

It was during this pre-takeoff period that public discussion and conjecture reached its highest pitch. Many and varied were the reactions. My mother of course condemned it out of hand because it involved "pot-shooting" while the birds were still in the water. She wished him no luck. My dad worried more about the seaworthiness of the craft in rough water, and how Uncle Norman could find his way back in fog without a compass when he was already restricted to peephole visibility. Old Louis Anderson, who probably had more practical experience with wild geese than anyone, said unequivocally that the scheme would never work. He said that these geese had "very strong noses" and could detect the smell of a human many miles away. He pointed out that this feeble craft could only be paddled downwind, so it would carry the scent ahead of it and warn the geese. Uncle Norman's buckskin clothes would be anything but a help in this regard. No, unless he could paddle the thing up into the wind, the scheme would never work.

Bill Mace agreed with most of this, but added that if Uncle Norman were successful in paddling against the wind, he would work up such a sweat in that confined hut that you would be able to smell that wet buckskin at least a half mile *upwind* and that was further than his old gun could shoot.

There were some people who were not prepared to predict one way or the other, but did say that if he was right and it did work as well as he expected, the government should intervene and put a stop to it before all the geese in British Columbia were wiped out.

The day he picked was grey and calm, and the tide would be out to expose the reef by nightfall — the perfect situation. He departed early afternoon for his two-mile voyage. We watched him for several hours. To start with he tended to weave around and actually go in circles, but as he gained experience paddling through the slot, his performance improved. It took him about four hours to reach the outermost shoals where the geese were expected to congregate. This covered an area of about two miles. There were some geese ahead of him but none where he was. Towards dark he got where they were, but they were now where he first was. While we could see him, he was never within half a mile of an actual goose. Then it got dark and we could no longer see the geese or the floating boulder. We heard no shots. All was quiet, and then a gentle southeast breeze started to sigh through the trees and we wondered how he was doing. The wind increased during the night and by morning was blowing half a gale with rain. No sign of the floating boulder or Uncle Norman in any direction. We reasoned that he would probably have made it to Indian Point where he would moor his raft and walk home, but that evening we went down to his cottage to look for him. No Uncle Norman. The wind increased to a gale and there were no boats on Savary in those days that would venture out in that kind of weather. That night, however, during a brief lull in the rain squalls, we saw what appeared to be a flickering fire on the beach on Hernando Island.

The next morning the southeast wind had abated so Bill Mace and Louis Anderson fired up the old four-horsepower Fairbanks two-cycle in Louis' old boat the *Red Wing* and went pop-pop-pop all the way over to Hernando, where they found Uncle Norman, safe and sound but a bit uncomfortable. The floating boulder was away up in the beach logs and Uncle Norman was living inside it. He had been eating clams for the two days and they said the empty shells were piled up nearly as high as the boulder. So far as I know, the subject of the floating boulder was never discussed again.

Obviously the floating boulder was not the reason for the disappearance of the brant geese flocks, so where are they? And where are the kind of people who spent so much of their time and effort chasing them?

FIVE

Swallow Hard If You Feel Something Hairy

IN THE EARLY YEARS ON SAVARY it was just about impossible for Dad. He'd gone to Repton School in Derbyshire with the man then president of the Hudson's Bay Company, who came through Vancouver once while Dad happened to be in town and put in a word with the local manager. Dad was overdrawn and overdrawn again, and the Hudson's Bay would carry him. If they hadn't, he never could have made it. He still had a bit of hilltop property down at Whonnock and he was trying to sell lots but it was slow going. He ended up selling much of it to Japanese berry farmers. It wasn't considered farmable, but they'd take this high land, clear it, and put in a little vegetable patch. They didn't use horses to plough, they had a thing like a mattock with a crooked handle and a wooden blade like an adze with a metal tip on it, and the women would get out there humped over all day long—chop, chop, chop—roots and everything, tilling the soil. They grew the most marvellous gardens, and strawberries like the white man couldn't touch. These people were being brought in by an agent named Takahashi and Dad sold acre after acre of this top land. Then he had to try and collect from Takahashi and I remember him talking to the lawyer about that. It never came through.

Dad was absolutely no businessman. He could never collect a debt, never ask for money, never assert himself. Mother was the assertive one, and she had quite a talent for selling. She was great for talking things up, all gung ho to go. She had a lot of ideas about what I should be doing, what Dad should be doing, how we should be carrying on and all the rest of it. She had more ideas than splinters in a woodpecker, but she was not much for following through. She'd ride off in all directions at once. She had some money of her own, but we never knew how much. She was secretive about it. She talked in thousands and yet never seemed to have any money when it was needed. She was a very selfish person, really, just the very opposite of Dad. He was very generous and easy going. He'd give in to her every time.

Mother had to be making a show of her independence at all times. She loved to be the centre of attention. Without fail, if a trip was being organized, she'd wait until the last minute and then back out. Just to have everybody falling over her. She loved that. Her way of dressing served her very well in getting attention too. She had very nice hair but in protest against something or other she had it all clipped off boy-style. That's the way she always wore it from then on, and she always wore pants. Sometimes knickers, with puttees or high leather boots. It was sensational for a woman in those days. She was talked about for miles around. Once when she was down in Vancouver she went to the Jericho Tennis Club for a game with this old English codger she knew, and they threw her out of the ladies washroom. They thought she was an interloper from the gents' side! She regarded herself as a man. She hated women. She used the word *females* in a way just dripping with disdain that meant the lowest of the low. Everybody we knew admired the hell out of her, but I hated to be around her. I would get so embarrassed.

It was a queer upbringing. Damn queer.

After I finished school Mother's plan was to send me to sea. She loved uniforms and was all for getting me into one. She loved the army, she loved the navy, she loved the cricket team, she loved them all. Dad wasn't especially high on the military and had no use for army types. They were stupid, stuck-up—he had everything against them. But he was favourable to the navy. As a kid, when he got out of private school, he had tried to get into the Royal Navy but couldn't because of his age. He was eleven. And the problem with that, to the Royal Navy way of looking at it, was that he was over the hill. They wanted you as soon as you were weaned. Their absolute limit was nine. Dad was bitterly disappointed and nursed a feeling all his life he'd missed his calling. He loved to talk about the navy. I picked it up from him, so of course I was all navy too.

There was no way of my getting in the Canadian Navy in 1919 but my mother came up with the notion of putting me into the merchant marine. This brainwave was helped along by the fact one of our summertime neighbours, Mr. R.R. Burns, had a brother who was the head of the Canadian Robert Dollar Steamship Company. The company had never taken apprentices, but Mother got busy, she went to town, she got in to see this guy in his big offices there, and she talked him into changing the company's policy. The company started taking apprentices and I was number one. The deal was no pay the first two years, $15 a month the third year and $25 a month the fourth year, by which time I should write my second mate's papers.

The Dollar company had sawmills and large camps over at Union Bay on the Island and also up Burrard Inlet — that's how come the name Dollarton. They were in the lumber export business too, with the *Melville Dollar*, the *Albert Dollar*, the *Bessie Dollar* — about seven or eight ships, all hauling BC lumber to the Orient. They were a rather chintzy type of operation, a small branch of a big San Francisco shipping company operating out of Canada on the cheap and not at all set up to support an apprenticeship program. I was assigned to the *Melville Dollar*, due to sail from Vancouver in two weeks.

When Mother came home with this news it was already set up. She had paid a deposit. There was no backing out. Not knowing any better, I was overcome by the glamour. Dad had no say in the matter.

There was much rushing around and preparation. We had to go to Vancouver a week ahead so Mother could buy me clothes and I was taken around to say farewell to all our friends and get their good wishes. First of all, however, Mother was very eager to get me fitted out in a uniform, so she took me to a tailor who had many customers in the merchant marine and knew exactly what was needed. Mother was delighted with the result and had me wear it at every possible occasion, although she was very concerned that the Canadian Dollar Company didn't have an official cap badge.

We stayed at the old Alcazar Hotel in Vancouver while awaiting arrival of the ship, but the company received a wireless message to the effect that the ship had been disabled some days out of Vancouver by an engine room explosion which had killed six crew. We waited two weeks before she finally limped into port.

For those days the *Melville Dollar* was a reasonably modern vessel. A 10,000-tonner built in Glasgow in 1906, she was 370 feet

Mother had more ideas than splinters in a woodpecker.

"Brassbound and copper-bottomed," October, 1919.

long, had four masts and two well decks with bridgedeck and funnel amidships. She had three Scotch boilers and a 2,500-horsepower triple-expansion steam engine that gave her a maximum cruising speed of 14 knots at 72 rpm. Her captain, W.W. Wright, was the youngest in the service at 44 but was nevertheless an old sailing ship hand and tough as nails. I later saw him knock three Chinese crewmen to the deck with his fists for what he termed "cheek." Except the four deck officers, four engineers and the purser, the crew she carried was all Chinese. Counting me there were 64 altogether, and I didn't really fit with any group. I was forbidden to speak to any officer or engineer unless first spoken to. The crew I could talk at to my heart's content, but it had to be in Chinese if I wanted an answer. Only one, the chief steward, spoke English. In other respects I was luckier. I was assigned a cabin all to myself. It was on the main deck opening into the dining saloon and was undoubtedly intended for a passenger. It had upper and lower bunks and a porthole which provided very welcome ventilation on warm tropic nights.

The eating arrangements were interesting, but probably standard for that type of vessel. The main dining saloon on the main deck was adjoined by the pantry and scullery. The pantry crew consisted of the chief steward, a waiter, and a dishwasher. The galley was also on the main deck but on the starboard side aft the funnel, separated from the saloon by about a hundred feet of open deck. All the food had to be carried forward to the saloon by a waiter who needed good training so he could make a dash in between waves in heavy weather. This was bad enough, but on our Pacific crossing we took on a deck load of coal piled eight feet high behind timber bulkheads. Until the pile was used up, all deck traffic had to climb a ladder, run over the coal, and then down the other side. This included the waiters with their trays of food.

The main saloon was just for the captain, the three officers, the chief engineer, the purser and myself, the "fourth officer." The three Scottish engineers had their own lunchroom amidships adjoining their quarters. The chief engineer said he would rather have eaten with his own crew, because it was closer to the galley and the food was hotter. The 52 Chinese were divided into two distinct groups housed in the forecastle right in the bows of the ship. The forecastle was divided down the centre, with the deck crew on the starboard side and the "black crew" (engine room and stokehold) on the port side. The two groups didn't mingle. I never saw them speak to each other. Each did their own cooking on a very small coal stove with a makeshift chimney poked through one of the deck ventilators. From the bridge you could see the smoke coming out of the cowling. Sometimes the wind would change direction, blowing the smoke back down into the forecastle, and you would see them come tearing up on deck to switch it around. They also kept pigs in the spaces between the big stacks of squared timbers we carried on deck. You'd see them running around squealing and copulating while being doused by waves.

For days on end when we were bucking into head seas water coming over the bows would go down the ventilators and put the Chinese crew's fires out. Once a big sea tore the whole cowling away, leaving them with no means of cooking or heating water. I remember going in and finding water dripping down in all directions and sloshing around on the deck. To keep their bedding at least partly dry they had wrapped oilcloth around their bunks.

One old friend of Dad's in Ruskin who had spent twenty years of his life "before the mast" on old square-rigged sailing ships, told me before I set out, "Be sure to call everyone *sir!*"

I remembered this, and felt I was well prepared. On my first day at sea the captain sent for me to come to his cabin so he could acquaint me with the rules: where I would sleep, where I would eat, to whom I would report. I was very careful to say, "Yes, sir!" to all these instructions. When he finished he said "Very well, boy, you can go now. But as I was going out the door he called after me, "And one more thing, boy. It's *sir* when you're addressing me, goddammit!"

I said, "Yes, *sir!*" and scurried down the bridge ladder.

My official schedule was:

> 6 A.M. to 8 A.M.: on the bridge including a trick at the wheel
> 8 A.M. to 12 noon: work on deck under chief officer cleaning, scrubbing, painting
> 12 A.M. to 4 P.M.: navigational studies under second officer.
> 4 P.M. to 6 P.M.: on the bridge

I kept a diary which graphically describes my experience on that first crossing:

Oct. 19: left Vancouver
Oct. 20: started regular duties but very seasick. No meals.
Oct. 21: still sick
Oct. 22: still sick
Oct. 23: still sick
Oct. 24: feeling a little better but no meals.
Oct. 25: storm from S.E. Sick again
Oct. 26: still sick
Oct. 27: Captain administered vile tonic called "Black Draught." Very giddy.
Oct. 31: Still sick and very weak.
Nov. 1: same
Nov. 3: same

I remember one day I was at my usual post, hanging over the rail losing my breakfast, when Jock, the fourth engineer, came along. He patted me on the shoulder, looked concerned, and said with his thick burr, "Kid, I want ya to promise me—if ya get a ticklin' feelin' like somethin' hairrry in your throat, for chrissake, swallow hard. Tha's your arrrrsehole!"

I smile now but I didn't then.

We sighted Hokkaido on November 7 and the next day passed through the Straits of Tsugaru into calmer waters in the Sea of Japan. We were soon out of sight of land again, although a strong smell of sulfur pervaded. This was from some unseen volcano. I started taking regular meals but I remained very weak. Before our anchor was down in Karatsu, our first stop, we were surrounded by Japanese "bum-boats" and peddlers were throwing grappling hooks over our rails and clambering aboard. I wasn't interested in the jewelry but I had always loved Japanese oranges and after three weeks off my food the scent of them drove me crazy. For fifty cents I got all I could carry to my stateroom. They were tree-ripened and quite different from the way we see them. You didn't have to peel them—they just fell out of the skin. They were unbelievably delicious. I don't know how many I ate before I stopped for a breath, but afterwards I daren't get over ten feet from the head. I was stopped up with three weeks' accumulation of black draught, and to say it cleaned me out is an understatement. I felt like a new kid.

Karatsu was just a coaling port. You could see mines from the water. The coal came down chutes into sampans propelled by long sweeps trailing from the sterns. They got up quite a good turn of speed, two men to a sweep, and came alongside the ship while still anchored out in the harbour. They would suspend a series of platforms like a stepladder from our railings, with about six women on each step, and set up a steady flow of 25-pound baskets from the sampan, up the ship's side to a smooth hardwood chute leading down our hatch. Young boys gathered up the empty baskets in stacks ten deep and ran to the rail, dropping them back overside. With sampans going on both sides of the *Melville Dollar* we had our bunkers filled in less than twelve hours. It was very efficient and impressive, if a trifle labour-intensive.

On November 14, out of sight of land, we entered a buoyed channel about a mile wide in very shallow brown water. This was the mouth of the Yangtse River. One of the officers told me that the CPR *Empress of Asia* had to wait for high tide to get through, and when she was put in drydock her bottom was found to be polished shiny from the sand in the Yangtse. Going up, weeds and rice paddies appeared on either side, and eventually a few flimsy structures. The channel was very busy with traffic from all over the world. Several Blue Funnel Line steamers passed looking very clean and smart, making us conscious of our rusty sides, peeling white paint, and grey, salt-encrusted funnel. In one place we had to anchor for several hours waiting for a thick fog to clear. All the traffic did. While anchored all the ships kept their bells ringing, surrounding us with a weird symphony. Then came the throaty yelp of a steam siren down river.

"Got to be navy," said one of the officers. Sure enough, with another deafening blast of the siren, the HMS *Hawkins*, a light cruiser, loomed out of the fog and streaked past upriver, crewmen dangling from bosun's chairs nonchalantly redoing her tropical paint as they went. It was one of the most modern naval vessels afloat, launched just after the armistice. I was thrilled to be in the same river. Then the fog lifted and we were under way again. We continued upstream all day.

We left the Yangtse and proceeded up the Hwang Pu to Shanghai. We secured to a mooring buoy and unloaded into lighters the first day, then went alongside the company docks just above the main part of town. In doing so we lost our steam steering gear and nearly went up on the bank, scattering junks in all directions. There were hundreds of them, all with patched brown slatted sails, carved upper works, and all with an eye painted near the bow, even the tiniest row boats.

SS Melville Dollar *with a deckload of timber.*

Coaling up with wicker baskets in Karatsu.

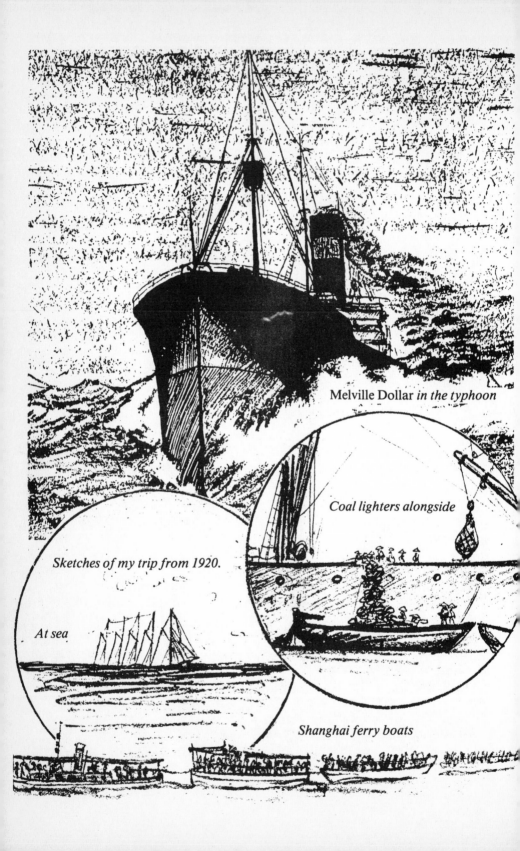

Melville Dollar *in the typhoon*

Coal lighters alongside

Sketches of my trip from 1920.

At sea

Shanghai ferry boats

The river water was very dirty and full of every kind of flotsam. We saw a concentration of sea birds busy on something in mid-river. As the current brought it past, I was shocked to find it was two bloated corpses somehow tied together and floating face down. The third officer shrugged and said they were just victims of pirates up the river. I noticed that the larger trading junks working the river had old-fashioned muzzle-loading cannon mounted in the gap at the bow as a caution against pirates. There were gunboats patrolling the river but the pirates had them outnumbered.

As soon as we got in to the company dock, we started unloading our deck load of timbers. These were squared fir logs 10 by 12 and 16 by 20 and about 40 feet long. The ship's windlass would lower them down onto two-by-four skids. They weighed tons, the large ones. Barefoot coolies would swarm around them and get rope slings underneath, two men to a sling. There would usually be about fifty coolies to make the initial lift, each with a bamboo stick across his bare shoulders, then a strawboss in black silk would stroll along in pointed slippers tapping any he considered surplus on the shoulder with his cane until there were just enough to bear it away. They moved with a shuffling step and sang "Ho-ho, ho-ho" to keep step.

It was "Ho-ho, ho-ho" all day long from a thousand throats.

They worked a twelve-hour day around the clock. Every six hours the paymaster came and paid them off in silver coins. They would take the money over to street vendors to buy food, mostly boiled rice. There wouldn't be much pay left by the end of the day.

At one place on the dock there was a large pile of rusty iron wheelbarrows. Apparently a few years earlier the company had been reclaiming some land behind the wharves and old Captain Dollar, the founder and owner of the shipping company, decided he would introduce some American efficiency. The Chinese were going about it in the traditional way, with hundreds of coolies shuffling along carrying baskets suspended from either end of a four-foot bamboo yoke—ho-ho, ho-ho. Dollar decided to switch the coolies over to wheelbarrows, and had his Yankee bosses carefully show them how to fill these wheelbarrows and wheel them along. The coolies smiled broadly, filled their barrows—then got a man on each side, put a sling underneath, lifted the barrow off the ground and shuffled away—ho-ho, ho-ho...

Incidentally, I saw old Robert Dollar while we were there. He came aboard momentarily and frightened all the officers out of their wits. He looked exactly like the Uncle Sam character the cartoonists like to draw and had the reputation of being a godawful cheapskate—always on the lookout for some little sign of waste.

The weather in Shanghai was hot in the day but frosty at night, more like interior weather than coastal. One of the best things about being there was the relief it gave us from our usual diet of putrid meat and weevil-filled bread. In port we ate a lot of fresh fruit, including a delicious pear-shaped grapefruit called pomelo. On the 20th of November, 1919, we cleared Shanghai and on the 25th reached Hong Kong, where we tied to a mooring buoy in Kowloon Harbour. The captain sent for me and informed me he was now going to fulfill a promise to my mother by taking me ashore, and I should get on some decent clothes. I had no clothes at all except those I'd been swabbing the decks in and my uniform, which I had been carefully preserving for just such an occasion. I gathered myself into it and appeared before the master.

He was aghast.

"Where do you think you're going like that, all brass-bound and copper-bottomed?" he bellowed at me. He refused to be seen with me unless I wore something "proper," but I had nothing else so he left without me. No one on the ship, it turned out, even possessed a uniform, much less appeared in one in public. After the captain was well out of sight the old chief engineer took pity on me and he and the purser took me ashore for a couple of hours. That was the only time I set foot off the ship for the entire voyage. Boy, would I get rid of that goddamn uniform when I got home! In the meantime I would use it for all the dirtiest jobs I could find.

We departed Hong Kong on November 30 and arrived in Manila December 3. I enthusiastically noted a large wireless station with high antenna towers. We were taking on bales of cigar tobacco, which gave the ship a wonderful aroma. When one bale dropped accidentally and broke open on the deck the officers and crew helped themselves to large bundles of leaves. I gathered this particular accident occurred every trip without fail. I grabbed a bagful for Dad.

The bales were brought alongside in native lighters which were about 75 feet long and had whole multi-generational families living in small cabins at the back. They all seemed very jolly, jabbering away in Tagalog and laughing, and in the centre of each cabin they suspended a huge handmade cigar. It must have been a foot long, just swinging there on the end of a cord. From time to time one of them, sometimes a wizened old woman and sometimes a small child, would grab the cigar and take a great big drag. It smelled just wonderful. It was the finest Manila cigar tobacco, world renowned. They would roll one on their bare thigh then wet it with their tongue. It just took a minute and they were set for the whole day.

On December 5 we left Manila and steamed for Shanghai with strong winds and a typhoon warning on the wireless. I, of course, became once again very seasick. The captain changed course to run in behind Formosa out of the typhoon's path, but it didn't work. On the 9th our stern log showed we were making only 1.5 knots against the fierce seas. On the 10th our speed dropped to less than a knot and soundings showed very shallow water. The ship's generator broke down, leaving us without lights, refrigeration or wireless. The next day we actually lost ground as the furious seas drove us backward over the log line and fouled it in the propeller. The next day we made a little headway but the leadline showed we were practically aground. We saw a lighthouse on the Chinese coast but had no idea where we were. The captain, sweating blood, changed course to seaward, looking for deeper water. On the 13th we sighted the sun for the first time in seven days and were finally able to shoot our position.

When we reported into Shanghai the next day, we found ourselves listed among fourteen vessels missing in the typhoon. Having been nine days without a report from us on a trip that was only supposed to take three days, they had given up hope. We departed Shanghai on December 17 and arrived in Karatsu on the 20th, taking on coal alongside a Russian whaling ship with dead whales aboard and two tame bears running around on deck. Very unpleasant to leeward.

Two days later we arrived in the Shimonoseki Straits where we took on a Japanese pilot for the Inland Sea, but apparently he didn't know his business and after a frightful row the captain put him off in a passing fishboat. Without a pilot we couldn't legally navigate the Inland Sea so we had to swing out around the island of Shikoku in a manoeuvre rather like going up the outside of Vancouver Island instead of through the Inside Passage.

Our steering winch had now irreparably broken down and we were reduced to steering by hand, which required four strong men heaving for all they were worth on two large wooden wheels aft, getting directions via hand signals from the bridge. The steering task was further complicated by the presence of a great many small fishing boats. Their mere presence was not the problem so much as a peculiar superstition they all seemed to hold: they believed that if they crossed close under the bows of a large ship their luck would improve. We would see them from a distance and steer a course to avoid them, but they would quickly haul in their lines, start their one-lung engines, and set on a collision course with us, scraping across our bow at the last instant. We were doing about twelve knots and were completely helpless to turn or stop at that point. Then they would stop so close they would practically rub down our sides and quickly heave out their lines before the magic started to fade.

This drove the captain into a rage because if we hit a Japanese vessel there was no saying what the Japanese government might do to us, but it would be disastrous. He sent down for a bucket of coal, specifying lumps the size of grapefruit, and each time one of these boats went under he'd rush out to the wing of the bridge and bomb them with coal and curses. He was too excitable to have good aim, so it didn't have much effect, and finally with one particularly wild swing his gold ring flew off into the sea.

We thought he'd been as mad as he could get before, but when that happened we realized he'd only had his throttle half open.

On December 25 in Kobe waiting for repairs we had a bit of a Christmas party but nobody was really in the mood and the Chinese attempt at plum pudding left something to be desired. We got away on the 29th and kept land in sight to port all day as I practised taking a type of cross bearing called a running fix. I had learned how to use the sextant to get azimuth bearings by this time and John Clayton, the second officer, said I would know enough to get my second mate's papers by next trip, although I would still need the four years sea time before I could sit for them.

Four years! The thought of three and a half more years of this life staggered me. I couldn't imagine it. It seemed like I'd been away from home for more years than that already. I had managed to fend off seasickness since the typhoon, but it had been replaced by a homesickness almost equally severe.

No one did more to try and divert me from my troubles than the chief steward. He was large for a Chinese, and probably in his sixties, but it was hard to tell. He spoke and wrote English, and even used a typewriter. He taught me a bit of Chinese, including how to count. He found out I liked to draw and dug out a Chinese pen, which is like a pointed brush, along with rice paper and a block of dry ink you used like watercolour. I spent some time doing black-and-white sketches of places and ships we'd seen. This made quite a hit. Soon everyone on the ship wanted a sketch. With the old steward acting as interpreter I learned that the bosun wanted me to paint a picture of the *Melville Dollar* on the inside of his sea-chest lid. Some of the officers asked for sketches to include in their letters home. It got my mind off my loneliness and succeeded in making me feel a part of things for a bit, but the picture that kept coming into my mind was the shining shore of Savary Island. I had never even thought about it before as a place, but now it seemed so obviously superior to any other part of the world I mourned over how simply I'd been tricked into giving it up.

Four years!

The second officer, John Every-Clayton, who hailed from Cumberland, BC, turned out to be a distant relative, with Derbyshire connections, and was as nice to me as his position would allow. Much later he became harbourmaster of the Port of New Westminster, where I did radio work for him. Andy Anderson, the purser and wireless operator, was very considerate and not at all stuffy as far as his position was concerned. He would let me sneak into the wireless room while he was on watch, but if the captain ever came around I'd have to ditch it and hide. I wasn't supposed to be in there or in the engine room, but I spent most of my time in one place or the other.

I was, of course, very eager to know more about the engine room explosion that had caused the delay on the last trip, but it was still very fresh in their minds and no one was very eager to talk about it. I finally got the details from one of the engineers.

Steam is supplied to the engine through a six-inch diameter pipe about four feet long that comes through the boiler-room bulkhead in a continuous right-angle bend and connects to the throttle valve. During operation the main engine, which is about twenty feet high, will vibrate from side to side as much as half an inch. This flexing between the engine and boiler is absorbed by the pipe. Apparently after fourteen years the metal fatigued and blew apart under 225 pounds of steam pressure. The engine room and stokehold were instantly filled with superheated live steam, killing three stokers and two oilers. No engineer was in the room at the time. At that temperature you couldn't even see the steam. They would be cooked through before they knew what happened.

The main steam shutoff was in the boiler room so there was nothing the crew could do but wait till the fires were out. The vessel drifted helplessly for several days. The chief engineer volunteered to be lowered down through the skylight of the boiler room in an attempt to reach the shutoff valve, but he was badly scalded and had to be hospitalized. The engineer's description of handling the cooked bodies wasn't pretty. They managed to repair the burst pipe by wrapping it with asbestos packing and sheets of copper cinched down with U-bolts, and that's how they limped into Vancouver. They installed a new pipe, but kept the old one aboard as a spare—or maybe as a reminder. It certainly gave me a lifelong respect for live steam.

January 3 we hit bad weather from the northeast with snow flurries. It got so cold the captain gave up his unheated cabin under the bridge and moved into the purser's cabin, which was over the engine room. For ten days we bucked strong headwinds and snow. We had over six inches on deck at one point. Breaking seas carried away our deck load of coal and all the wooden bulkheads along with it. Then we ran low on coal. The boilers could burn either coal or oil, whichever the company could buy cheaper, and there was a part tank of oil left so we switched to that. It soon became apparent that we would run out of both coal and oil before reaching Vancouver so they got the crew out on deck with a crosscut saw and cut up all the available pieces of timber lying around. Then they stripped all the shelving and flooring out of the ship's storerooms. When that ran out we were still well out of Vancouver, and they began searching madly around for anything burnable they might have overlooked. The only thing there was lots of was paint. Black for the hull, white for the topsides, and red for the funnel; there were thousands of gallons. Somebody got the idea—paint burns, let's try that. They dumped a couple of drums into the oil tank, it burnt, so they dumped in the rest. We staggered into Vancouver Harbour burning straight paint. Another day and we would have been dead in the water.

The officers were pretty disgusted. They said there was no reason for it. I detected some of them taking a certain vengeful delight in the destruction of the ship's woodwork. We had lots of bunker space, but between the cheapness of the company and the poor planning of the captain they just didn't get enough fuel to see us home. Meanwhile the unnecessary cost of rebuilding the storerooms and resupplying the paint stores would dwarf any savings the company gained through its famous stinginess.

It wasn't until we arrived that I realized I had made the entire crossing without once succumbing to *mal de mer*. I decided that after the experience of the typhoon I was proof against anything.

Mother and Dad came to Vancouver to meet me. It seemed a lifetime since I'd seen them, and I felt I was now quite grown up. For Mother's sake I put on the remnants of my uniform. She looked me over with great pride but wanted to know why I didn't have a cap badge yet. Oh, God!

I asked them aboard to my cabin where we could talk. Then I dropped the bomb. I had decided to leave the ship. I was finished. I was at pains to have them realize this was a mature decision and nothing to do with the fact I was seasick for much of the voyage. That unseamanlike weakness was all behind me, and my reasons were on a higher plane. All the officers, even the dear old Chinese steward, agreed: the seafaring life was a dog's life, and I should get clear of it while I still could. They had backed me, bringing it up with the captain, and he agreed to recommend to the company that no financial penalty apply, since it was a first for them and they had a lot to learn about training apprentices.

Mother was terribly disappointed, but was half expecting it from some of the letters I had written. Dad was hugely relieved and there were tears in his eyes when he hugged me.

We packed my kit bag and went down to the Union Steamship Company to board the old SS *Cheakamus* for Savary Island. About one hour out of Vancouver we ran into a light squamish wind blowing out of Howe Sound. The *Cheakamus* started gently rolling and wouldn't you know it, right in the middle of recounting my grand experiences I threw up on the deck before I could get to the railing! I was humiliated, but it probably did more to convince my mother than any of my earnest arguments.

That's as close as I ever came to fulfilling my grandfather's wish by making it big in the shipping industry, but in a roundabout way that trip did end up providing me with my first career. I went back to odd-jobbing when I got back to Savary, but my plan was to save enough money to learn radio and return to sea as a ship's wireless operator.

Going to sea had done some other things for me, not all good. It had taught me respect for my seniors — and every other person on board was senior to me. This required addressing everybody else as *sir* in each and every spoken sentence. And I was strictly forbidden to speak unless first spoken to. This kind of respect and subservience was not expected back home among the loggers and fishermen of the BC coast. In fact, it was despised, but it was so ingrained in me that it took months to break the habit.

My vocabulary had expanded also. Nautical language tends

toward the colourful—nothing like old Louis Anderson when he was fighting a hangup, but colourful just the same. For instance, the adjective *bloody* was applied to everything, and occurred three or four times in every sentence. I adapted to this readily, having decided in my own mind that it was perfectly clean and harmless. But you have no idea the kerfuffle that was caused when I got home and tried it on my mother—very respectfully of course—such as "Pass the bloody butter please." Or "How's the soup dear?" "Bloody good, thanks!" Both my mother and father, who had never been heard to use anything stronger than "Damnation!" were properly shocked by all this, and made it the subject of a very serious lecture, which embarrassed me. After all, I *was* fifteen, wasn't I, and just retired as "fourth officer" of the SS *Melville Dollar!*

SIX

Engineer from Fourth Reader

MY PLANS FOR A CAREER IN WIRELESS couldn't go anywhere without money, so the first order of business on returning to Savary was to find gainful employment. As it happened, early in 1920 a new logging show turned up on Savary in the form of a log float accommodating a cookhouse, a bunkhouse, and a ten-by-twelve Empire steam donkey, plus a small blacksmith shop and other odds and ends that went to make up the bare necessities of a logging operation of that day. The owner was a small but very determined man named Walter Prescott. Prescott was very tough physically, but only recently a logger, having originally been a prairie farmer and road building contractor. Mrs. Prescott was a very quiet and refined lady and quite a socialite. A very unusual combination, but it seemed to work. She lived in a very tiny private cabin, also on the log float, and had pink curtains in the windows. Prescott had a crew of about three men, including an engineer whose sole experience had been running a high-wheel steam tractor and threshing machine on the prairies. The other men referred to him as a "goddamn hayburner," but he did a good job and kept the old machine from falling apart. If any highball donkey-puncher had got his hands on it, the operation would have been terminal.

This was one of the first ten-by-twelve donkeys ever built by Empire Machine Works in Vancouver. It was distinguished by having a cast iron main frame which had not stood up to the weaving of a donkey sled and had huge patches bolting it together in two places. Nevertheless, this was by far the largest and most powerful logging donkey ever to make an appearance on Savary Island. Prescott located his camp right at the Springs, tied the float up to the nearest tree stumps, and moved the Empire straight up the hill and then along the bank about a quarter of a mile. The spring was dammed up and a pipe was run straight into the cookhouse, and everything was ready to go in two days.

Prescott let it be known that he could use some help if anyone wanted a job, but at that time Bill Mace had several new houses to build and anyone available on the island, including my dad, was working for him as carpenters. But I was fifteen and raring to go, and there were two others, Barlow and Fritz Tait. Fritz was my age and small. He got a job as whistlepunk. Barlow was seventeen and taller, and he was taken on as chokerman. I was a swamper and knotter. The only thing that wasn't very clear was when we would get paid, because Mr. Prescott was fresh out of money, but he said as soon as the first boom was sold, we would all get our money. How many times has this story been peddled in the logging industry? He paid us a flat 25 cents an hour, somewhat less than a full-grown man would get, but to us it looked good. Eight hours a day, six days a week — twelve dollars was not to be sneezed at in those days.

It was about three months before the first boom was sold, but all the money was needed for groceries for the cookhouse and we had to wait for the second boom. When this was sold and a cheque arrived in the mail, Harry Keefer, making the most of his powers as small-town postmaster, arbitrarily held it up until the outraged Prescott agreed to come up with enough cash money to pay us half of what we were owed. He offered to increase our wages to 40 cents an hour if we would stay on, but as far as I now remember, I think it was a case of "Once bitten, twice shy." But it was not all bad. He fed us well, we learned a lot, and we enjoyed every minute of it.

Old Prescott wanted to do everything right up to date. Ground yarding was not for him. He had seen some high lead shows up the coast, and they really moved the logs, so he was determined to do the same. The first thing he needed was for someone to climb about a hundred feet up a large fir tree and cut the top off where it was around thirty inches in diameter. He had acquired an old set of spurs and a climbing belt in anticipation of this. He told Barlow, with the explicitness of one whose authority is beyond question, to take the spurs and go up the tree with a saw and axe, and chop the top out of it.

Barlow was even more explicit. He told Prescott he could shove the whole goddamn tree up his ass and twist it. He wasn't having any part of it. Prescott tried everyone else but there were no takers.

"What a shitless bunch of bastards," the old man scoffed. "Gimme them goddamn spurs. I'll go up 'er myself!"

For his age, he was amazing. He stripped right down to his heavy Stanfield underwear and caulk boots and huffed and puffed his way all the way to the hundred-foot mark, chopped the top out, then climbed up a second time with a pass rope to rig the guylines and high lead block. He certainly won the respect of the crew. After that we would do almost anything he asked, providing it wasn't to climb that tree.

The setup was relatively simple but worked quite well. It certainly was much better than ground-yarding. With the lift provided by the hundred-foot spartree, there were far fewer hangups. After all the logs within reach had been piled at the tree we swung the mainline around the tree and took it down to a big rock on the beach. The choker with the log connected to it was hung from a bicycle block which ran up and down the main line when it was tightened. This is known as the gravity tight-line method of dumping logs into the water. When the donkey pulls the mainline tight, the log is lifted clear of the ground by the bicycle block and is dragged screaming down the hill by gravity, taking the haulback line with it so the bicycle block can be hauled back up the hill after the log is unhooked from the choker.

Unhooking was Barlow's job. It was not too bad when the tide was out and the logs were lying on the beach, but this was inclined to break a lot of timber and so they usually chose to dump at high tide. This meant that Barlow had to scramble out on the floating logs to reach the choker and, having no caulk boots, he would frequently fall in. My job was to help the hooktender choke the logs in the pile and then run to the top of the hill where I could see Barlow and give the engineer hand signals. Since we only used flat open-face choker hooks in those days, they often unhooked themselves when the engineer slacked the line down, but you couldn't count on this, even though the flathook was notorious for coming unhooked when you didn't want it to, during yarding. At that time we had never even heard of a Peterson hook, with its locking knob and bell setup.

This tightline system, as haywire as it was—including a homemade bicycle block—worked surprisingly well for the most part, but there was one weakness, and that was that the old donkey had a very ineffective brake on the haulback drum, so the engineer had no control over the load once it started flying down the hill. In fact, the brake would hardly stop the drum from spinning when the load reached the bottom. The engineer would usually swing the friction lever over to help the braking action, a little like putting your car in reverse then easing out the clutch to brake going down a hill, but this wore out the oak friction blocks and was not popular with Prescott.

I vividly remember one impressive screw-up. It was towards the end of the operation. The hooktender had quit, so the engineer was doing the hooking-on, with my assistance, and old man Prescott was running the donkey, a job for which he was inclined to be too excitable to do well. We had just hooked on a fat log. We waved to the old man to "take her up," which he did with much snorting and clouds of steam, and over the edge it went, hell bent for the beach, with the haulback just screaming off the drum, a situation calling for close attention by an alert engineer to watch for the first appearance of slack in the haulback, whereupon he should pounce on the brake.

The old man didn't pounce quite fast enough. A loop of loose cable flipped over the end of the drum and all hell broke loose. The load was halfway down the hill, going like a runaway express train, when the loop cinched around the shaft. The haulback broke at the worn part near the butt-rigging, and the freed end whipped back up the hill like Paul Bunyan's slingshot, shooting up and past the top of the spartree and cracking the whip about three hundred feet in the air with a noise like a cannon, then snapping around among the guylines several times before coming to rest. We had instantly taken cover under logs, and only came crawling out when we thought all was quiet.

There was still plenty of action around the donkey. The haulback drum was still spinning at high speed and was spewing out enormous coils of three-quarter-inch cable all over the mainline drum, covering the front part of the sled and surrounding ground with a tangled mass of quivering cable. Prescott was jumping up and down on the haulback brake, screaming, "Jeesus — Jeesus — Jeesus," but it was doing no good. The first flying loop of loose line had snagged the brake band right off the drum and threw it in the bush. He finally thought to slam the friction lever on and brought the whole thing to a stop in a great cloud of red rust. But what a godawful mess that was to clean up. Over a thousand feet of cable had to be untangled and then pulled out onto the landing, coiled and re-coiled several times, and then spooled back on the drum. The brake band had to be

Climber going up a tree.

South Shore of Savary Island, *a pastel I did in 1924.*

Mount Denman from Prideaux Haven as I painted it years later.

repaired and reinstalled. And then someone had to go up the tree to rerun the haulback.

There were still no volunteers for this job, so it was up to the old man to do it again. This time it should have been a simple matter, since there was now a passline and a pass-block hung at the top of the tree. One end of the passline had a short crosspiece of wood called an ass-bench on which the man sat, while the other end was taken over to the donkey and wrapped around the capstan on the end of the haulback shaft. As I said, it should have been easy, but in this case the passline had been made up from several short pieces of old five-sixteenths strawline, joined together with a lumpy cat's paw knot at each joint. Everything was fine going up. Barlow was tending the capstan and holding tension while the old man was working up at the haulback block.

When he finished his job, the old man waved down and shouted, "Slack 'er down easy, Barlow." Barlow let a foot or so of slack, expecting to see Prescott start moving down, but nothing happened. Prescott yelled again, "Slack, Barlow!" and this time Barlow let three or four more feet go, but still no action up the tree. The old man figured Barlow was asleep or something and got mad and yelled, "Barlow, you asshole, I said *slack!* Whatsamatter with you?" While shouting he started jumping up and down on the ass-bench like a mad kid. That's when the knot that was jammed in the pass-block popped through, and with all the slack Barlow had been dutifully paying out, the old man plunged a good twenty feet down the tree, bounced back up, and turned over upside down on the ass-bench. Barlow lowered him gently the rest of the way to the ground, expecting to be fired on his arrival, but the old man never said a word about it.

My great friend among the Savary Island bunch was Jimmy Anderson, a cousin of the large Palmer family who had most recently logged up Homfray Creek in Homfray Channel. This was one of the first examples of high lead logging in BC. Loggers came from all over the coast to see how they did it. They had four machines, four steam donkeys, and they logged about a mile and a half up the valley with skylines and tightline. When they finished up their claim in there and pulled out for new territory down in Theodosia Arm, old Louis Anderson got the idea that us kids should go in and cut shingle bolts for the summer. There was big cedar growing up that valley, beautiful long-grain cedar you could split boards out of. Cedar like that, if you don't fall it onto a mattress made of limbs it splits all to pieces, and on this high lead show with the rocky ground they had broken an awful lot of it. They left it all laying, so his idea was Jimmy could take a couple of chums up there and saw it into shingle bolts and bring them out. In those days there was a good market for shingle bolts. They're normally 54 inches long, as I seem to remember, and about as big around as a kid can easily reach. Jimmy approached me. I was really hungry for a way to start earning money, so Jimmy, Jimmy's little brother Terry, and myself set out to spend the summer making our fortune cutting shingle bolts.

Old Louis Anderson was going to take us up in the *Red Wing* but Jimmy's cousin George Palmer came along with his speedboat. George was the high rigger in the family and a bit of a high roller in a BC coast kind of way, and he'd built what all of us considered to be a very exciting boat. He'd hollowed out a big log and put a six-cylinder Buick motor in it. The log was flat on the bottom and he hadn't got it quite straight, it had a hog in it. This made it very fast. But it also had a bit of twist in it, so the faster you went, the more it heeled over, until you were sure it was going to flip itself upside down and dump you in the chuck. The whole time you were thundering along in that thing every muscle in your body was unconsciously straining to keep it upright. Anyway, George had come down from the new Palmer-Owen camp in Theodosia on his way back to the old campsite at Homfray Creek to get something, and he said he'd take us up. They had left a lot of stuff up there, including all the buildings, one of which we were going to live in. We loaded in a whole lot of stuff, loaded some more into my own twelve-foot kicker, and set out across Desolation Sound in this very fast log, towing the kicker.

We got as far as Prideaux Haven and the log sheared its shaft off. George was a good mechanic but there was no fixing it, so I got my little 1½-horsepower Evinrude wound up and towed the whole kaboodle into the nearest place we could get help, which was Archie Stewart's on Redonda Island. Archie was an old handlogger who had a beat-up cod boat with a four-horsepower Easthope in it and George's idea was to get a tow into Lund for repairs. But Archie's boat wasn't working either. The Easthope had a broken coupling. Well, George *could* fix that, but old Archie didn't want any gyppo logger with a broke-down log to touch his precious Easthope. Like a lot of old-time handloggers, he considered motors next door to black magic and was convinced that only Frank Osborne, the machinist at Lund, could properly minister to it, so he'd been waiting for someone to come along and tow *him* to Lund.

Archie was just one of the great old characters who lived around Desolation Sound in those days. Across in Melanie Cove there was Mike Shuttler, the philosophical hermit Capi Blanchet was so impressed with in her book, *The Curve of Time*, and down at Portage Cove there was old Joe Copeland, who used to meet the steamboat dressed in a full Confederate Army uniform, complete with the little cap that kind of folded in the middle. Old Joe was a bugler on the Confederate side and his Dad was a colonel on the Union side. After the war he was in one of those renegade gangs that went around robbing stagecoaches, and finally escaped up into Canada. He ended up here in this little neck of land where the Gifford Peninsula joins the mainland, trapping and handlogging. The Copeland Islands, which form part of what locals call the Ragged Islands just north of Lund, are named after this old Civil War outlaw.

Archie Stewart was the more serious logger and he had two claims, a beach claim and one up the sidehill. He beavered away all by himself, cutting the trees down one at a time and laboriously peeling the bark off on one side, then rolling them over onto the peeled side with his jackscrew and sliding them downhill to the water. One time they noticed he hadn't been into the store at Refuge Cove for several weeks so they went over to his shack, found no sign of him, looked around, couldn't find anything, his boat was there, and so they finally started following the fresh skidmarks up to where he was working on his summer claim. There he was under a log, three weeks gone. They just piled some rocks around him. Archie Stewart's still up there.

He was a stubborn old cuss and on this day didn't especially want to lose time from his valuable enterprises helping us, but eventually George convinced him they could haywire his boat together, and they unloaded all our gear onto Archie's dock and he and George set out for Lund towing the defunct log. We three boys were left to ferry our summer's supply of tools, saws, bedding, clothing, and grub the three miles or so across Homfray Channel in my little kicker just as darkness was coming on. The three of us went in with a small load first off, and it was pitch dark before we were halfway there. I couldn't see anything. I didn't have the faintest notion where we were. I was vaguely aware of high mountains towering up around us.

Jimmy had a mischievous streak in him and he couldn't resist rubbing it in a bit.

"Oh, my goodness me, the wolves are bad here. Grizzlies use this place to come down out of the mountains. Cougars too—they hide up on the branches waiting to drop down if you go by—shush! I thought I heard a wolf howl. . ."

Jimmy was a wonderful character. He had a natural talent for imitating things. He had wonderful powers of description. I don't know where he got it from. He was limited in his words, but if he didn't have a word he'd invent one, or invent a noise, and when he was done it would be indelibly imprinted on your mind. I remember him telling me what it was like to fire a gun on top of a mountain.

"Gee whiz, Jimmy," he said (he always called me Jimmy). "It sure was funny. I was up that mountain with Uncle Bill, you know that one up Toba? We were up on the top and my, it was clear. All you could see was other mountain tops. There was another mountain top over there, it must have been a mile away, and you could see a goat walking around on it. Bill, he got up his rifle, adjusted the sight, he pulled the trigger, it went click, and it was like this: the gun went *puh*. You could hardly hear it. I looked at Uncle Bill and he looked at me and for fifteen seconds we didn't know if it went off or not. Then: HAR—ROOOOOOOOOOOM! My goodness me, the noise almost knocked us over into Toba Inlet. And then HAR—ROOM, HAR—ROOM, HAR—ROOM. It seemed to go on for ten minutes. All the echoes from all those mountains."

He was only one year older than me but he knew his way around Desolation Sound a lot better. I was amazed—he just steered me on a beeline through the pitch dark night and before I knew it we came to a notch in the shore and we lit our coal-oil lantern and clambered up the beach, and there we were at the cabin of the former camp timekeeper, which Jimmy had earlier decided we would use for our summer headquarters. We left Terry in the cabin to lighten the load in the kicker while Jimmy and I went back for the second trip.

With the little putt-putt it took a good two hours pretty well, all in the dark. We got back to Archie Stewart's, stowed another load, putted back, landed in the boulders, carried the stuff up to the house, went to open the door—and it was barred. Jimmy yelled, "Open up. Let us in!"

No answer. We both yelled, and the more we yelled the less we heard. We hammered on the door, pounded on the walls, and eventually had to break a window. We found Terry in a corner completely buried under a pile of blankets. After listening to Jimmy's stories he'd been absolutely scared stiff, and when he heard all this racket he figured his time had come.

Old man Anderson had worked out a pretty slick method for getting the bolts down off the hill. He sent down to Vancouver and bought a thousand feet of quarter-inch black iron wire. Pretty husky stuff. Came in a big spool. We'd tie one end to a stump across one side of the creek, then run the other end up the opposite bank through the area we were cutting bolts in, take it over a strong branch, and tie it down to another stump. Then we'd tighten it with a jack screw until this thousand-foot line was singing tight. Threaded on the wire were half a dozen cast iron pulleys with wires hanging down and dogs on the end. Each pulley setup ran on the wire like a trolley. One guy would grab a dog wire, the other two would hold up a shingle bolt, and the first guy would drive in the dog. Then we'd let it go and zhiiiinnNNGG, it would go flying down the line, bash into the low-end tailholt stump and fall off in the creek bottom.

We'd send six bolts down, then while Jimmy and I cut six more, Terry would trot down and fetch the pulleys back. The idea was to cut up all the cedar for a couple thousand feet up the banks on either side and fill the creek bed with our cut bolts. This was summer and the water was low, and when the fall rains came, the creek would rise and wash all our booty down to salt water. The local name for Homfray Creek was Roaring Creek, and it could usually be counted on to build up considerable force as it plunged down the gulley during wet weather.

We rustled around on the sidehill sawing up cedar slabs and pulling wire all summer. Old Louis Anderson would come up once every two weeks in his old *Redwing*, bringing more supplies from Woodward's. He'd buy one trip and Dad would buy the next. This went on until August and we had the creek full. The old man said, "You can't do any more until the creek floods now, there's no use going on." We put a hanging boom of old cedar chunks tied with wire all around the outlet of the creek to catch our bolts in case they started washing down before we got back. We reckoned we had over a hundred cord and we expected to get about sixteen dollars a cord from the shingle mill down at Redonda Bay. I remember gleefully adding it up on paper, like Midas counting his gold. Terry and I went back to Savary to see all the friends we'd missed that summer and catch up on the fun, and Jimmy stayed behind to watch camp.

Suddenly there was a change in the weather and the old man said,

"There'll be enough water in the creek now, you boys better go back up." We piled into the *Redwing* and chugged off in the direction of Homfray Creek. When we got as far as Prideaux Haven we met shingle bolts. They were all over Homfray Channel. There'd been a thunderstorm in the mountains and Homfray Creek had turned into Roaring Creek in a matter of hours. It became an instant torrent. The force of the water was such, Jimmy said, the shingle bolts were going under our hanging boom and over it. We spent the next two days crisscrossing Homfray Channel in the *Redwing* dogging runaway shinglebolts. We gathered up all we could find in the area, brought them all back in, got a pocket boom around them, and ended up with about eighty cords. Then we had to get them down to Redonda Bay on the west end of Redonda Island, where the mill was. Redonda Bay was then called Deceit Bay and it was a thriving place, with a big railway logging camp, a cannery with a large fleet of fishing boats, a store, and the shingle mill. In later times it became a wilderness prison camp and you couldn't go ashore there. The mill manager in those days was Sid Vicary and Jimmy had arranged with him to take our bolts at the going rate, depending on how good they were. There were three grades. He would send up a tug when the time came.

We got word to him and sat tight waiting for the tug to arrive. When it came, it was a little steel towboat called the *Topaz*. I don't think it was 25 feet long. Steam. Originally it was built for construction work on the Panama canal, and had found its way up the coast. I don't imagine it would be ten horsepower. They hooked onto our makeshift boom, which was about as hopeless a thing to tow as you could imagine, requiring very gentle handling—and very calm weather.

They reached Deceit Bay with about half of it. They hit a squall coming down Toba Inlet. Meanwhile Sid Vicary announced the price had gone from sixteen down to eight dollars a cord, and we'd ended up with only about thirty cords. It just paid for the groceries.

That was my first business venture. It couldn't have filled the ghosts of James Ward and the Governor with much hope, but it taught me something about the dangers of working things out on paper.

It was a well-spent summer just the same. I'd lived on Savary Island for years and looked up Desolation Sound at those mountains every day of my life but I'd never been close to them. I had no concept what it would be like to reach out and touch the mountains. Savary is so flat. Hernando is so flat. Lund is rocky but not high. In my little kicker we managed to get as far as Manson's Landing over on Cortes Island, which is flat, but even that was a long way. We never could have got up through Desolation Sound and into Homfray Channel on our own. It's wonderful country, despite the gloomy name given it by Captain Vancouver. Very beautiful protected waters walled around with spectacular mile-high peaks — Mount Whieldon, Mount Aitken, the Unwin Range, Dudley Cone, Mount Denman just behind us on Homfray Creek, and across on Little Redonda, Mount Addenbroke. Little Redonda Island is one of the tallest, smallest islands you'll ever see. It sits in front of Toba Inlet like a castle with high walls, hollowed out inside by Pendrell Sound, a very deep inlet that is one of the coast's geographical oddities due to the exceedingly warm water there. It is the best place in BC for getting Japanese oysters to spawn. On the east side of the island is another geographical oddity, where Mount Addenbroke, which is exactly a mile high, falls in a sheer drop into the deepest sounding on the continental shelf — 401 fathoms. This is the highest land nearest the deepest water on the continent. Opposite Redonda on the mainland shore of Desolation Sound you have Prideaux Haven, a charming little jumble of islets and lagoons which has now become the coast's most popular marine park. You can walk across it on fiberglass from June to September, but in those days Desolation Sound was almost as desolate as when Vancouver named it. When I woke in the morning in that place and the mountains were roaring, with these godawful thunderstorms, I felt like I'd landed on a different planet. It was a frightening place.

We hadn't quite gone broke on our great shinglebolting venture, and our summer was far from being a dead loss to me because I learned enough about scrambling up sidehills pulling line and swinging a crosscut saw to become fairly useful around a logging camp. I began to get work with Jimmy at his uncles' camp down Theodosia Arm. I did well in camp, by golly. After a few months I started running donkey. I still wasn't old enough for it but I faked my age and got my Logging Donkey Engineer's steam ticket. You were supposed to be 21 and I was only 17, but I got away with it. The paper I got permitted me to write L.D.E. after my name — for Logging Donkey Engineer. It wasn't quite the P.Eng. or Ph.D. Dad had once hoped for me, but it was an engineer of a kind. Once or twice I've found my name being listed alongside a bunch of radio or aviation pioneers or landscape painters who have lots of degrees behind their names, and my L.D.E. has helped even things up.

Working alongside Jimmy Anderson was always an adventure. His mother was one of the Palmers, among the oldest families in that area, and she was terribly ambitious for her children. She had three boys and two girls, Terry Anderson, Andy Anderson, Jim Anderson, Sylvia Anderson, Pearl Anderson, and then much later two more, Lloyd and Judy. Mrs. Anderson wanted all her kids to be president of the United States. They all had the same schooling as I did, or less. Andy planned to become a civil engineer but the poor guy died of pneumonia. Terry was extremely dull—he never learned anything. He wasn't even a good logger. Jim was a tremendous person and spent the rest of his days logging. That's the only thing he could do and he had the knack of doing everything wrong and yet not killing himself. I could write a book about all the trouble he got into in the woods and survived. In the short time I was in the woods there were three or four occasions he should have been killed right while I was there, but he never was.

One time I'll tell you about—Jimmy was running donkey and a runaway log came down the hill straight for the donkey. The chaser saw it and yelled a warning. A steam donkey, with its great mass and all its warm places to crawl, was a very secure and hospitable place to be up to a point, and that point was precisely where the general level of mayhem around the landing became a threat to the security of the boiler. When that point was reached the standard plan of action was to get as far away as possible as fast as possible, because a burst boiler could cook everybody for a hundred feet around.

Jimmy hears the warning, looks up at the charging log, and decides to clear out. He's jumping down off the sleigh onto the ground when the woodbuck hears the warning and decides the only safe place is behind the boiler, up on the donkey. They meet in mid air and fall down in a tangle of arms and legs. The log comes up over the front bunk like an express train, jumps the drums and makes a direct hit on the boiler. When I get there the boiler is lying 250 feet down the hill looking like you'd stepped on a beer can. Nobody is hurt.

Jimmy had a real peculiarity, and that was that he didn't swear. I don't know why, he had so many opportunities. He never smoked, he never drank, he never swore. His only curse was "My goodness

The Palmer-Owen floatcamp up Theodosia Arm where I started logging in 1922.

R & R at Palmer-Owen. Don't be fooled by the little skirts—this is one tough bunch of bush-apes. Jimmy Anderson with arms folded.

Going ahead on the haulback, the most underage logging donkey engineer in BC perches at the controls of an 11-by-14 Washington.

me." When I got up to see this donkey wreck, he was telling me all about it.

"My goodness me," he said. "My goodness, Jimmy, you know that boiler fizzed and spurted steam for a good hour afterwards."

One other time I was punching donkey and we were working at the top of the chute at Theodosia Arm. They had a 1,400-foot gravity chute from the top of the hill to the chuck and five donkeys strung out from there on up the mountainside. It was an Empire steam donkey with a tree on the brow of the hill. George Palmer was the high rigger for the camp. They were a well-balanced family. They had specialists for everything. George was the climber.

Why he lived, I don't know. He was careless.

On this occasion the haulback block on the top of the spartree was starting to squeal. It needed greasing or it would burn up. Normally that was the high rigger's job, but George wasn't around, so Jimmy said he'd go up the tree and take care of it. Jimmy at this time wore Harold Lloyd glasses, the kind with big, perfectly round lenses. They gave him a surprised look. He didn't need glasses at all but his mother thought they made him look studious. He could barely read or write, but I guess she figured the look would be a start. He kept them for many years. Any time he wanted to see anything he'd always pull them halfway down his nose and peer over.

For taking a man up the tree they had the same arrangement as Prescott had—a light line called a passline, which was threaded through a pulley at the very top of the spar and was tied back out of the way during normal yarding operations. It had the same short length of two-by-four shackled on the end and you'd get this between your legs and the donkey puncher would very slowly go ahead on the line, taking you up the tree. Jimmy got this ass-bench under him and somebody gave me the signal to go ahead easy on the passline. From my position running the donkey I couldn't see Jimmy because the brow of the hill cut off my line of vision. The first thing I could see was about twenty feet up the tree.

I started ka-chuff ka-chuff ka-chuff with the passline lever down and pretty soon I saw Jimmy coming up, looking pleased with himself and signalling me to put on some speed. He got up past the buckle guys—there are two sets of guylines holding up a wooden spar, the buckle guys about halfway up, fanning out in a radial pattern, and perhaps forty feet above that the top guys, also fanning out in a radial pattern to tailholt stumps around the landing. In between these two sets of guylines you had the block Jimmy was going to grease. The passline which was pulling Jimmy up the spartree led through the pass block, which was the highest thing on the tree, up above the top guys. This pass block was a light block which was hinged to drop open and release the line but was held shut during use by a pin locked in place with a cotter key.

Jimmy cleared the buckle guys. He was swinging around, as a high rigger will, because the line under tension will tend to untwist. I remember seeing him stick his boot out against the tree to stop the spin. I must have looked down to see how the line was piling on the drum. There is a danger the line, as it winds in, will pile up on one side and fall over, subjecting the rigger to a nasty snap, and when you've got a man's life dangling on the end of a string ten storeys up, you want to be extra careful.

I felt a tug on the line and looked up just in time to see Jimmy disappear. The line went slack and then snapped tight again. I looked up again and Jimmy appeared again, but about fifty feet over to one side of the tree—and he was upside down. And then he disappeared again. And then the line tugged again. Then he appeared even further off to one side, turned a complete somersault, and dropped out of sight.

What had happened was George, the careless high rigger, hadn't bothered to put a cotter key in the pin that held the pass block shut. It had gradually worked its way out, and when it did, the block dropped open. The passline fell out, and Jimmy dropped down twenty feet to where the passline hung up over a top guy. When this happened, it whipped him up in the air again. This happened two or three times until he'd worked his way far enough down the sloping guy to hit the ground on his feet.

Well, I thought we'd killed him.

I ran up there and here's Jimmy—kneeling on the ground with his head down as if mortally wounded.

"Jimmy! Is everything okay?"

"My goodness me, no!" he says. "I lost my glasses."

I can date this story to the hour because shortly after we broke for lunch and walked down to the cookhouse. Being the first machine up the hill we were only a ten-minute walk from camp so we could go down for a hot lunch. We were all sitting down when suddenly everything got very dark, indoors and out. Anyone with a biblical turn of mind might have assumed Judgment Day had come. As it was, nobody said a thing. We didn't know what was going on because we had no way of getting news at Theodosia Arm then, but

September 10, 1923 is recorded as the date of a rare full eclipse of the sun. In a few minutes it got light again, and through the clouds we could actually see the brownish moon partially covering the sun.

Jimmy let out a sigh of relief.

"My goodness me, I was worried for a minute," he said. "I thought it was from that bump I got."

I stayed with Palmer and Owen until they finished up in Theodosia, then went over to Squirrel Cove and logged all in behind Squirrel Cove and down to Von Donop Creek. I made good money there, oh boy. A licensed engineer has to get steam up — a fireman can't do it. A fireman can only fire if there's a licensed engineer looking after the boiler. So I had to go up an hour ahead of the crew. I was getting seven dollars a day for a nine-hour day. I had been in the woods less than two years and I was already on top of the logger's world. But radio was still the thing I wanted to be in, and as soon as I had a thousand dollars saved, I quit and went back to Savary.

SEVEN

Voice from the Sky

IT'S EASY TO PINPOINT THE TIME I first became aware of wireless: April 15, 1912. I was six years old, and Dad and I had taken the local train from Whonnock down to New Westminster to see somebody about something, I think it was the H.H. Heaps Lumber Company office we went into, and everybody was standing around waiting for the latest telegrams. The *Titanic* had just sunk. It developed she had sent out this call for help via wireless telegraph, and this made a real impression on me. The idea a ship sinking in the middle of the ocean could somehow send out its call for help all that distance over nothing just really intrigued me.

The next encounter I had with it was going over to Victoria with Mother and Dad on the old CPR steamer. I was very young but I do remember on this trip going past in front of the parliament buildings — in a horse and carriage. Of course you can do that again today, but this was not a tourist affair with pneumatic tires, this was an original working outfit. But the important memory of that trip comes from the voyage back on the old steamer. Near the purser's office was the wireless office. The CPR was equipped with wireless then. My mother couldn't get me away from the door.

The operator was sending off traffic reports or something. To Point Grey in those days. What he was using, as I know now, was a big rotary spark gap, probably the old one-kilowatt deal. They had a bank of Leyden jars and this spark-gap — this was a pinwheel rotating with electrodes just almost touching but not quite, and the current would jump that and spark. They drove it at five hundred cycles a second—a nice tone to copy. The receivers were just a crystal and a cat's whisker and a tuning coil. And a pair of headphones. It had a range, with a good antenna, of fifty miles or more. It would cover Vancouver to Victoria. There was VAB in Vancouver and VAK in Victoria and the ship would be VGRB or something like that. Anyway, I was standing outside the door and the operator was using the key, sending messages out over the ether, and I couldn't pull myself away. From the start it had a special attraction. The ozone smell from the sparks came to be the smell of adventure for me.

Then when I went to sea on the *Melville Dollar*, here it was equipped with the latest in Marconi wireless. It was like the one I'd seen on the CPR boat, but even more modern—it had, in addition to the crystal detector, a vacuum tube receiver. There were six what they called Marconi valves. They were about four inches long and had a brilliant filament that lit up like a hundred-watt light bulb. They took the incoming signal and amplified it through a resistance-coupled amplifier—then a horn speaker. They were very insensitive. To control the volume there was a rheostat on the bottom and the lights dimmed down or they brightened up.

That was the equipment on the *Melville Dollar* and through the kindness of Andy Anderson the operator, whose nickname was naturally Sparks, I became quite familiar with it. Sparks wasn't able to really explain very much to me. He knew a little about batteries and so on, but I probably knew more of the theory than he did at that point. Somewhere down the line I'd read one of Dad's books from Cambridge about electricity and got the idea that if I couldn't be a civil engineer, then how about an electrical engineer? They were just starting to talk about electrical engineers and I didn't even think you had to go to university. When I got back from sea I really had the bug and I bought the ten volumes of Audel's electrical guides and dictionary. A dollar apiece and two dollars for the dictionary.

My chum on Savary, Jimmy Anderson, also became involved in wireless, mostly at the prodding of his mother. She was determined he would become a radio operator. God knows why. In 1920 radio was far from being a household word. There was no voice broadcasting and the use of wireless was limited to industrial applications and a few hams, all using Morse code. Mrs. Anderson shelled out $150 for a correspondence course from the Radio Institute. I can see the ads yet. The guy in the picture had a long, thin face you wouldn't trust anywhere. They sent you what they called an omnigraph. It was like a hand-wound phonograph with a disk you put on and it had a contact and a buzzer and one single headphone. It would send letters and numbers in code and you could run it any speed you wanted. But it only gave you about a minute's copy, then you had to buy more records.

That's when I decided to build a telegraph line, from my house to Jimmy's. He lived down the other end of the settlement a little over two thousand feet from me. I remember the distance because I had to order the wire. I got galvanized fence wire. It only cost about four dollars. I put cut-off beerbottle tops on wooden pegs on the trees. Some of them are still there, do you know that? Over half a century later. From the old tent, west of the government wharf, down to Jimmy Anderson's place, east of the wharf. Then I made relays out of old Model T generator cutouts. I probably got directions from some magazine article. We hooked up a buzzer, and for power we went over to Frank Osborne's machine shop in Lund and got old Number Six dry cells thrown out by the fishermen, who used them to run the spark coils for their Easthope boat engines. You could cut a lime-juice bottle off by tying a kerosene-soaked string around the neck, lighting it afire, then dumping it in water—cuts a bottle right off. I did hundreds of them, because I used the necks for insulators. The bottom I used to put the dry cell in. Then I bought about ten pounds of sal-ammoniac. I don't know what sal-ammoniac was used for, but Frank Osborne had lots of it around his machine shop and it was very cheap. These old dry cells would be really dry by the time we got them, all bubbly around the bottom, and we'd punch a bunch of holes into the sand inside it with an icepick and set it down in this cut-off bottle in a solution of sal-ammoniac and water. After a while that soaked in, livened up what was left of the zinc, and the god-darn cell worked again. Not for high output, but I had boxes full of these things all wired up. That got us our power.

I suppose I picked many of these tricks up from Frank Osborne, whose machine shop over in Lund had already become one of my favourite hangouts. Frank took a great interest in my various crackpot schemes and I can't touch on any story of my early years up the coast without bringing him into it. My earliest recollections of Frank were based mainly on his manner and appearance. He was reasonably tall but stooped, moved slowly and deliberately and seldom spoke except to answer in a monosyllable or a grunt. He was stooped, I guess out of habit, as he was always bending over a lathe or a drill press or a forge in his machine shop in Lund. He never moved quickly or without deliberate decision because he spent most of his waking hours in close proximity to operating machinery and moving belts and spinning flywheels. The fact that he retained all his limbs, fingers and toes, had both eyes and bore no visible scars attested to his "think first" attitude.

His manner sometimes appeared rude to those who didn't know him. To those who knew him well, he was an extremely kindly man, just not easy to get excited or even to get an immediate response from. Once again, this was partly due to his vocation. He would be sitting crouched over a lathe with belts flapping, smoke spiralling off the lathe tool, and long silvery metal turnings dropping to the floor, when some anxious individual with a boat engine that was needing repair would enter the shop and go and stand expectantly beside him trying to get his attention. Frank would not take his eyes off his work for perhaps five or ten minutes. When the cut was finished, he would reach up and release the chuck, take the work out of the lathe and throw it on the floor, spit a long, accurately directed stream of tobacco juice down through a square ten-by-ten hole in the floor about ten feet away, turn slowly around to the individual, look over the top of his glasses at him, and grunt, "What do *you* want?"

Whatever the man's problem, providing it was a mechanical one, Frank would invariably take care of it, but only in his own good time and after the customer had come to the full realization that he'd be looked after when his turn came. If the man was patient and kept quiet, Frank would stop what he was doing and attend to the man's requirements on a first priority basis. If he liked you, nothing was too much trouble. I learned this lesson very early on and went out of my way to be respectful and polite, and over the years I got the feeling that Frank treated me more like a son, probably because he had no children of his own. I respected the relationship. As a kid I loved nothing better than an excuse to spend time in the machine shop watching Frank work, and whenever there was an occasion, talking to him.

There was a certain omnipotence to Frank's presence which was enhanced by his personal appearance. He always wore dark clothes. They may originally have been dark blue overall material, but were always so saturated with oil, grease, and coal dust that the effect was black. He wore a very old, battered, greasy black felt hat. He was bald, but this didn't show as he always wore the hat. His face was generally as black and greasy as his clothes. If he ever did remove his hat, which he had the good manners to do in the presence of a lady, I was always surprised to see how white his head was. His shop was very poorly lit, and you could be standing quite close to Frank and not notice him. Only the whites of his eyes would show up until you got accustomed to the gloom. He did have electric lights, but they were very small and concentrated just over the working area of each machine. In the blacksmith shop part he said he wanted it dark so he could see the colour of the metal he was tempering in the forge. I got to love the smell of blacksmith's coal smoke mingled with the smell of the dogfish oil he used on all the cutting tools.

He still had all his teeth and all of them were perfect — and boasted that he had never been to a dentist in his life. He credited it to the fact that he had chewed tobacco continuously since he was ten years old.

While Frank had perfect teeth, remained in full possession of all his limbs and faculties, and enjoyed generally good health, he did require glasses for close-up work, and these were a constant source of trouble to him. He was always misplacing them or breaking the frames, or they would fall off his nose into the forge, and so on. Then one day he showed me how he had got fed up with these "goddamned opulists" and made his own. Actually he took the lenses out of a pair of broken frames he had. They were perfectly round, like Jimmy Anderson's. He got a piece of 1½-inch diameter brass pipe, put it in the lathe, and parted off two narrow rings. He machined grooves in the rings to fit the lenses and then brazed a brass nosepiece between the rings and bent it to fit his nose. To finish the job he brazed on a couple of wire loops where the hinges would normally be, and then attached a band of red inner tube about a half-inch wide which he snapped around the back of his head. They gave him an owlish look, but they worked fine. When he didn't need them, he just pushed them up on his forehead, where often enough they would be when he was hunting all over the shop to find them. He wore them for years. They never fell off and he never completely lost them, and they wouldn't break.

Once Frank told me about his early life and I wish now that I had paid more attention. Apparently he was an orphan, brought up in a monastery somewhere in the Ozarks or in that general hillbilly area of the United States. He worked on a farm and learned his blacksmith trade as a kid while at the monastery. Anyway, he wound up on the coast of British Columbia and settled down at Lund before the turn of the century, where he started a blacksmith shop and then a machine shop, both of which were badly needed in those days. He was a real creative genius, and very advanced for his time. He designed and built his own marine engine, the F.P. Osborne Heavy Duty Marine. He produced and sold two basic models — the Osborne 7 Horse Power Single Cylinder and the Osborne 14 Horse Power Two Cylinder. I seem to remember they bore a resemblance to the Atlas Imperial built in the United States at that time. In order to produce these he even operated his own foundry, right in the shop. I would guess that he probably ceased manufacturing them about 1925 or so. I can remember seeing them in quite a few of the local fishing boats. After that he invented, or at least designed, the first gasoline-powered boomstick borer, and these became widely used all over the coast.

With Frank's help we eventually got our telegraph line going and began practising the code. I learned the code, Jimmy learned the code, and Mrs. Anderson learned the code. Every night we'd sit down for an hour or so and talk backwards and forwards on our telegraph line until we got our code speed up. I helped Jim with his course. I was way ahead of this course at the start and I went right through it with him. Every lesson. When he got to the final exam, I wrote it for him. So he got the diploma but I got the knowledge. Jim never got anything out of it. That really helped me. When I got some money I sent away for a much more advanced correspondence course in electrical engineering from International Correspondence Schools for $165. They were red books. They issued them every so often, I had to do my homework and send it in, have it marked, and it was a real heavy deal.

My idea was to learn the theory, learn the code, and get my operator's licence so I could go on the ships. But Sparks talked me out of it. I wrote him saying I thought I would be making $120 a month. He wrote back, "I don't know where you heard that. I'll give you the present Marconi scale: you start at $55, the next year it goes to $65 and you end up at $110 a month after four years as chief operator." But he said, "Don't do it, kid." He and his buddy John Clayton undertook to discourage any notion of my returning to sea. As John Clayton said, "I been ten months at sea, and it's a year and a half since I been in the company of a woman." At my innocent age that hardly seemed like the worst thing that could befall a person, and anyway I had my doubts because I'd seen them in action in places like Shanghai and Hong Kong, making up for lost time. But they succeeded in undermining my ambition to a great degree.

I couldn't get through the correspondence course in any case. I came to the end of my mathematics. My fourth reader education just wasn't good enough to get me through what was really a university course. I got a third of the way in and I got a lot out of it but I had to give it up.

That was the end of my dream as far as the ships went, but not the end of my interest in radio. Some of the summer kids I grew up with on Savary—Walter Turnbull, the younger brother of Frank Turnbull the neurosurgeon, and Tom Barnett—used to get me down to visit in Vancouver during the winter, and in their basement they were playing with radio. It was all spark. They had a spark gap about a foot long, a wire out into the cherry tree, making a big smell of ozone, and listening on the crystal set—in those days there would probably be fifty or a hundred amateurs in Vancouver. Very few of them stayed with it. They were just kids going to school doing it as a hobby, then dropping it, because this was still before the radio boom, this was very underground.

My ambition was to go back to Savary, put a bunch of junk together, and talk to somebody not on the island. It was just at this time one of Jimmy Anderson's cousins, Norman Palmer, came along and gave me fifty dollars for a twelve-foot work boat I had beachcombed and installed with the Polish hermit's accursed two-horsepower engine. I spent it all sending away for radio parts. Montgomery Ward was advertising radio headphones, the cheapest I could find anywhere, and I sent away and got a pair of those. They were "The Lakeside Special." They were brown bakelite and cost fourteen dollars. I had them for years. Later you could buy them for two dollars a set. We used to take the Montreal *Star*, which was a very high-profile national paper at the time, and there was a column running for wireless buffs which told you how to make things. I got a crystal, made up a cat's whisker, and for the rest all I needed was bellwire and a circular oatmeal carton to wind it all on.

I made this thing and eventually I could hear boats. Morse Code. I could hear Estevan Point, I could hear Triangle Island—this was just before the isolation and fierce winds there caused the station to be abandoned in 1920—VAB was Vancouver, which is now VAI; VAC was Cape Lazo (now Comox); VAK was Victoria; VAE was Estevan; I don't remember what Triangle Island was. That wireless station was later moved to Bull Harbour.

So I couldn't go to sea and become a ship's radio operator but I could stay on Savary and listen to the deepsea ships coming in, talking to the shore stations giving their positions, their ETAs, and reeling off long lists of stuff. That normally goes at twenty words a minute and I finally got so I could copy it pretty solid.

I had weak reception but Jimmy climbed a tree for me and got a huge antenna up. I'd stay up half the night after everybody had gone to bed trying to tune in new places, and one night I couldn't believe my ears: I kept jiggling, jiggling, trying to bring it in and—yes, music! *Music!*

This was now 1922. We'd read about radio broadcasting starting up, it was all very strange and far away. I listened and listened and listened and finally I heard the announcer's voice: "This-is-K-P-O-San-Fran-cis-co. You are listening to Rudy Seager's orchestra in the Fairmont Hotel." I darn near died. I got Dad out of bed—he thought I was going crazy. He listened to it, and the next night everybody on the island, pretty near, was listening to it. You remember when the Sputnik went over and everybody was standing out at night looking up at the stars for the first time in forty years? Well, that was nothing. This was voice, the *human voice* coming out of space. Inconceivable. And music! Played in San Francisco and heard in the same instant on Savary Island. For sheer shock to your system the Sputnik wasn't in the same league.

That started the thing off properly then. Everybody wanted me to build them a crystal set. Old man Keefer gave me $25. He said, "Build me the best one you can get!" I got to know the Northern Electric peanut tube that you could run off dry cells and I started building sets and selling them, just like that. By 1924 I was building eight-tube superhetrodyne receivers. I built one for Dr. Lea and he took it to Vancouver with him.

Just overnight, radio boomed. And it carried me with it.

I had trouble getting money to buy parts, and this was the period when I would go out and work in logging camps for a few months to make some money. Then in 1924 I came back with a thousand dollars to really go into the business, and I often say I never worked a day since!

EIGHT

A Radio Expert is Born

IN NO TIME AT ALL people up the coast were sending their money all over the place buying the best sets they could get. They cost a lot of money in those days—over two hundred dollars. Few people earned that in a month. And they wouldn't work. They were designed only for local reception. Nobody was selling a set that would work miles and miles away. That's where I was able to keep on top of the whole radio situation on the coast. I introduced regeneration. It was tricky to handle, but you could increase the sensitivity of the detector thousands of times.

In 1923 I joined the North American Radio Relay League and I subscribed to their magazine, *QST*. I've taken it from 1923 to this day without missing an issue. This kept me abreast of everything that was happening in radio technology, and I read everything I could get my hands on. I was more on top of radio technology in 1924 than at any other time in life.

I remember Dr. Lea coming up to Savary and when he saw what I was doing he said, "My God, Jim, you should move to Vancouver. You don't know how badly we need radio men down there. With your knowledge you could make a fortune." That didn't appeal to me one bit. I was a country boy. I was afraid of streetcars. Electric lights, my God. Going to Vancouver was like going to another planet.

Anyway, what was happening on Savary Island was exciting enough for me. There was lots of business coming in. People got to hear about it and they'd come down in gasboats from Homfray Channel, from Teakerne Arm, from Refuge Cove—"Can you fix radios? Are you the guy we've been hearing about? Can you take a look at mine?" They'd come back in a week and get it. Just to hear Fibber McGee, you know. They'd come miles by slow boat. You couldn't have got these people out for anything else, an election, a funeral, a war—they would be too busy. But this little bit of music to lighten up their lives in these dark corners of the coast, once they knew about it they had to have it. I hung a sign out on the Ragged Islands where every boat going north had to see it.

Indians would come in. Loggers, towboaters would drop their tows and steam over. "Where's this expert? Our radio's busted." They'd bring in a six-tube what we called a TRF set. Tuned Radio Frequency. They were strictly for local use. In Vancouver you couldn't hear Victoria. They'd work fine in the showroom on Granville Street, but up the coast you'd get nothing. Except at night, when some very strong signals would come through once in a while. My sets would drag stations in from all over the goddamn continent. KDKA, Pittsburgh, Pennsylvania used to come in every night. Get the time signals from it. WOR, Fort Worth, Texas was a strong one. CFAC, I think it was, would come in with "W.W. Grant, Calgary, Canada." It came right over the Rockies.

This made a tremendous impression on these people. They'd tell their friends, and soon they'd be in with their radio. I had a real production line going. I'd take a set in and rebuild it, I'd give them back the same cabinet, but inside would be a different radio. My charges would be, depending on the set, around $25.

I was just thrilled. It wasn't donkey puncher's wages, but I didn't have to go logging to get it. To be able to make money staying home and fooling around at radio just seemed like getting it for free. It was like magic.

My family took heart. Here I was barely twenty and launched on a successful business. They had been a bit quiet when I flunked out of the merchant service, but now they let me know they had great confidence in what I was going to do. My aunts all wrote marvelling at how well I was getting on in this exciting new field. I was determined not to let them down, but even I didn't expect it to come together as fast as it did. I was quite prepared for this to take twenty years or so to get started. I was very patient at that time. Later I wasn't, but then I was.

I was still very concerned about Dad and spent the summers

helping him do work around the island. Then in the winter I'd get back to my radios. Things became much easier for Mother and Dad around this time. My grandmother died and in 1924 Dad received twenty thousand dollars from the settlement of the Findern estate. My mother's friend Miss Burpee had much earlier bought two of the best front lots on the island beside the wharf and kept them for us, so my father now paid her five hundred dollars each for these and we built what by Savary standards was a very comfortable home.

My mother lived in the new house less than a year before she moved out. After all those years in the tent, she couldn't stand living in a real house. She would cook our meals when we were working and then she wouldn't eat with us. She would sit there and smoke a cigarette and put her feet up on the table and watch us eat. She lived in a shed out back.

In 1926 Dad got a contract for building a government road for a thousand dollars a mile and bought a new Ford one-ton truck. I helped him chain out the proposed route—4.78 miles, so we got $4,780—and it just took that to cover wages and expenses. Dad got foreman's wages, $4.50 a day, and the rest of us got $3.20 a day. We rented the truck to them for $4.00 a day, which we thought was a pretty good rate. Dad did an excellent job, but he didn't get any real money out of it. By this time he and I had officially set up in business under the name Spilsbury and Son. S & S was diversified: our letterhead listed well-digging, electrical work, wood cutting, motor transportation, land clearing, and real estate sales.

Then in 1927 when the Royal Savary Hotel was built up at Indian Point, we got a lot of work that winter. The road was still impassable so if the tide was out we would walk the beach, which was shorter, Savary being crescent-shaped. We'd go up in the dark carrying a lantern—there were about six or eight of us—quit at five at night and walk all the way back with the lantern again. We built the hotel that way. It was run for years by Captain Ashworth, then by his son Bill, in the process becoming the island's most famous landmark. Bill was largely responsible for this notoriety through his non-stop promotion. He liked to make the claim that Savary was a geographical and meteorological anomaly, being located in the place where the north and south tidal streams meet, which gave it a special climate not unlike that of a south sea island. He would sometimes appear dressed in a grass skirt to support this theory, looking like Ichabod Crane in drag. The Ashworths liked to advertise the hotel as a luxury resort, emphasizing the *Royal* and claiming no end of

My first shingle en route to the Ragged Islands, where it hung for years.

The first Savary Inn ablaze in 1926.

Building the Savary Island Hotel.

Royal Savary's infamous intertidal golf course. Mother on left.

Harry Keefer, Savary Island's Mr. Everything, making fun of our truck in one its unhappy moments.

Excited Savary residents admiring my 1928 Ford, the island's first car.

Spilsbury and Son woodcutting operation. With the drag saw, splitting malls and truck we could move a lot of wood in a day.

Maintaining Savary roads with the latest equipment: a horse grader and a Cat Fifteen.

high-class attractions, such as a large golf course. People would come in expecting something like Pebble Beach, only to be shown to this small, rustic lodge with the owner's wife doing the cooking.

"Where's this large golf course?" the skeptical guest would ask. Captain Ashworth would look out the window at the full tide.

"I'll show it to you in a couple of hours," he would say.

The "golf course" was the beach. They laid out holes on the sand, but they had to wait until the tide was out. They also tried to construct a course at the Meadows, the natural clearing on the west side of the island where the keekwillie holes were. Once, Captain Ashworth almost set the island on fire trying to burn off the wild grass there. It looked a little more like a golf course than the beach did, but only a sidehill gouger could play it.

Some guests would leave in a huff, but those who stayed often fell in love with the place and returned year after year. Still, people always wondered why the Ashworths didn't just bill it as the rustic little country resort it really was. Bill Ashworth operated the lodge until the mid-eighties, when he burnt it down rather than attempt to bring it up to current government standards.

When it became apparent there were passengers to be carried from the steamboat dock up to the hotel, Dad and I decided to go into the taxi business, and I went to town with some of Dad's money to buy a suitable car. Allan Mace said, gee, he wanted a car too. He didn't have much money but anything I could get for a hundred, two hundred dollars, would I buy it for him? I went into Vancouver Motors. The salesman there was very darn reasonable. He said, "Well you want something that's not going to give you much trouble. Here's one that doesn't look too good but it's really in good sound shape. The engine's been overhauled..." and so on. And he was right. It was a good car. A 1928 Ford touring car with room for five passengers. Then I said, "Now I want a cheaper car for my friend," and he practically gave me one. He said, "Look, we got a lot of them around here," so I got a 1922 Model T touring car for Allan for fifty dollars. We shipped them both up on the same boat. Great excitement on the island. Allan had all kinds of fun with that damn thing of his — and immediately went into competition with me in the taxi business.

In 1931 I wanted something pretty decent in the way of a taxi, so I bought a seven-month-old Plymouth Sedan from Emil Gordon down in Powell River. God, the deals you could get from that guy! This car had been bought by some fisherman up at Lund who got drunk, turned it over and damaged the roof. It had been sort of straightened out, touched up with paint and the car was in almost new shape. I think he was asking seven hundred dollars for it, but we gave him an old player piano we got for helping around the hotel, plus an encyclopedia in a good bookcase Dad had around, plus a hundred dollars cash, and the Plymouth was ours. Later I took it over to Lund and kept it there so when I was going up and down the coast I could tie up at Lund and drive into Powell River, and later still I took it into Vancouver on the barge and used it there for years. Finally I traded it in on a used Studebaker. Then one of our technicians bought it back. He used it up badly, and years later when I was getting off the Vancouver Island ferry at Departure Bay and walking ashore, there among a lot of cars that had been left along the side of the road was my old Plymouth, abandoned. It stopped me. It had been an intimate part of my life for fourteen years. It had played a big part in my first love affair. For a minute I had an odd twinge of guilt, as though I should have done better by it somehow. Then I thought, sheesh, what foolishness! and walked on. But it's funny how many times I've wished I'd gone over for one last look and given it a pat or something. They don't last long enough to get that attached to any more.

While Dad still had his nest egg he bought that truck, which was over a thousand dollars, he bought the car—then, to top it off, we bought a Caterpillar 15 tractor. I did a lot of land clearing and road building with that and the Department of Highways loaned us a two-horse grader with the big arches and a seat up on top, two small wheels in front and two big ones in back, and that was marvellous. We towed it mostly with the Ford truck. It sat for years and years just above high-water mark rusting away, and they tell me now it's on display in some museum.

We sold the tractor eventually to Gordon Armstrong, a ham radio friend in Kelowna, for $1,000. I think it cost us $2,500, about. Right now it doesn't sound like very much money, but in those days it was a fortune. We worried about it, because this inheritance was all my parents had, and they were getting on.

I ended up wishing we'd spent more of it. The balance of the money stayed in the bank and it used to bug me. I think bank interest in those days was three percent. I was talking to a lot of the guys I'd known and they were all making all kinds of money—the stock market was red hot and everybody was speculating for all they were worth. I had a talk to Lyall Fraser of the Vancouver Reliance

Company, which had been started by Mr. and Mrs. Wootten of Savary, and he said, "Oh gee, your Dad should invest that in something. It's a shame." The house had only cost about four thousand to build and Dad had quite a bit left. He gave Fraser the whole bundle. "We'll diversify," Fraser said. "You don't want to have all your eggs in one basket." He got a thousand dollars worth of this, five thousand of that. One of them was Johnson National Storage, as I recall. Some things were paying 22 percent.

We were all smiles. Then came Black Friday. The very same year. All Dad had left was the Johnson Storage. It wasn't paying interest, but it still had a little value. Everything else was gone. I found a whole bunch of certificates in the basement a few years back and took them in to somebody. There still wasn't one of them that was worth a thing.

I always felt so damn responsible for that. I was young and keen and I was very envious of the city people who were rolling in wealth, investing in these stocks and things at twenty and thirty percent. I'm sure that Dad would never have done it if I hadn't suggested it.

Still, with the lots, the house, the equipment, and me with my radios — we were sitting pretty compared to where we'd been. There was some jealousy on the part of the other people, like the Maces and the Keefers and so on — they were more inclined to spend their money on fun. They'd go to town. Keefer had a trip every year. Of course, he was the storekeeper. To us that made him very wealthy, but looking back, I wonder how he took in enough to survive. We didn't put on any show, but in our island projects we invested solidly — the best damn house on the island, the best two lots, best tractor. I suppose Dad worried, but I never did. I was going out into the world and doing certain things — if it didn't work I was no further back, if it did work I'd make a little. I had nothing to lose. I never borrowed money, didn't get into debt, so there were never, at that time, any of the kind of financial worries that load you down, make you think of suicide and this sort of thing. Everything that came in was to the good.

NINE

Five Hundred a Month

THE DEPRESSION YEARS WERE IN MANY RESPECTS the most exciting and the happiest time of my life. All of a sudden I found myself on a financial level more or less even with all of these rich kids I had to look up to coming up and spending their holidays on Savary, obviously looking down on me because we had no money, we lived in a tent, and so on. There was always a big gap between me and these kids who came up; between them and any other permanents. We were country kids, and made very much to feel it. Not intentionally, maybe, but you couldn't help it among kids. So all of a sudden I found myself just as good as anybody else. It was a wonderful feeling.

I guess what made that more important for me was that, slow though I was to realize it, I was not totally devoid of interest in the other sex, and by this time I was able to meet what girls there were around on more or less even terms. Actually, I had been paired up with one of the summer kids, Glenys Glynes, in a vague sort of way for some time. The family used to come up every summer for years, I knew them well, we played together year after year after year. There was no hanky panky or anything but it was just a sort of foregone conclusion among ourselves that someday we'd probably get married. Nothing exciting. Very practical. We were in no hurry. That time would come.

But in the meantime there was a gal who came up with the Vancouver Art School. Colonel Herchmer's daughter Laurencia had been to the Vancouver Art School and she arranged for them to start an off-season camp for students to come up and sketch in the wilds and take classes at the Royal Savary Hotel. There would be twenty or thirty of them and I would have to taxi them around with all their stuff. They had some big names, Jock MacDonald and Group of Seven founder F.H. Varley among them. E.J. Hughes painted several well-known seascapes there as a student.

This girl came through with one of the first batches in 1932. She was about sixteen, she was vivacious, attractive, and I'd never run into anything like it before. I fell head over heels in love. I was simply starry-eyed about her. She was learning to be an artist, which gave me something to go on, because I had carried on my interest in drawing and painting since I was a boy and I could hold my own on a sketching trip with most of these kids. They'd all be there with their easels up, wearing the landscape out, godawful stuff. I was very interested in photography at this time and one day I went back to get some pictures of them with the old press camera I was using then. There was one old guy with a pointed white beard teaching photography and he said, "Come along with me and we'll do some shooting together." I stopped to get a fine big arbutus overhanging the water and he stopped me. "No, no. Don't waste film." Finally he had me taking some stepped-on crab shell or something. All very strange stuff I would never have thought of taking a picture of. He was some famous highbrow artist-photographer, but I've no idea now what his name was. It was quite a stimulus and I learned a lot of technique in pretty short order. I set up my own darkroom and got a better camera, an Ikonta B, and began manipulating the image during print-making. It got me off to a good start in photography, and for many years I worked at that instead of drawing. You see, I wanted to make a record of what I saw. That was my motivation in drawing and painting, and photography accomplished it that much better.

This girl I don't think could ever have made the grade as an artist, but at that time she was full of the romance of it. Her parents were wealthy and her uncle was a big name in provincial business circles who owned a famous yacht that still ranks as one of the all-time classics. She would arrive up on Savary aboard this yacht and knock everybody's eyes out. It was a mutual attraction, we both felt we had something going. She wanted me to come to town, but she had no idea the problems I was up against trying to make a buck. I did go to see her on one or two occasions and I was made to feel like a real country heel. Boy! She took me to see her smartass artistic friends and I felt like something out of Little Abner. I felt terrible. I was older, I was dressed funny and everything else.

She put it up to me bluntly one time. She said she'd talked to her uncle about it and her uncle had told her — she had parents, but her uncle meant everything — he told her he would never approve of her marrying anyone who had a lesser income than five hundred dollars a month. In those days that was astronomical.

But I was determined. If it's got to be five hundred a month, how in the hell am I going to make that? And soon! It was pretty well unthinkable. But I just had to get it. Suddenly the patient, easygoing approach was out the window. I started raising my prices and chasing every dollar I could see. But I knew I could never make it the way I was going. I was going to have to expand my business far beyond where it was. The words of Doctor Lea, urging me to move to Vancouver and make a fortune, came back to me, but the thought still gave me the shakes. I knew I wasn't ready for that. Vancouver was wired for AC power, and I had never graduated beyond DC batteries. AC scared me. I got the idea of going to Powell River and working for Emil Gordon. Powell River was on AC power, and I could break in slowly with it there. That'd be a step. And when I got a start there, then I'd move to Vancouver and as soon as I was making five hundred a month I'd go and propose to her.

Emil Gordon was a wonderful character. I learned a lot from him. He had a music store in the basement of the Rodmay Hotel and soon opened another one down at Westview. He was selling radios like crazy and he needed someone to fix them. I went to work for a hundred a month and worked eighteen hours a day. I got board for $1.20 a day, and what board it was! A room in the hotel and meals with the Gordons.

Emil was a great big tall guy, wore a brown derby hat on the back of his head, and he had hands a foot long. He was into everything. He cut quite a figure around Powell River for many years. Powell River was a total company town in these days, but he had nothing to do with the mill. He didn't like the mill, but they tolerated him because the people liked him. He sold cars, he sold anything. He was Salesman Sam. He would never walk away without making a deal. If he had to give it away — even if it cost him a hundred — he'd make a deal. And take anything in trade. He called me one time to come down to the wharf. He'd been to Texada Island selling radios and he had some stuff to unload. He had a tug and a scow. And the things

Delivering artists and students to the Royal Savary Hotel, 1932.

Wearing out the landscape.

My kind of art — my new 1931 Plymouth under a full moon.

This is art? Photo I took under Art School influence.

My dear friend and mentor, Frank Osborne.

The Mary, *my first radio boat, as she looked when I first got her.*

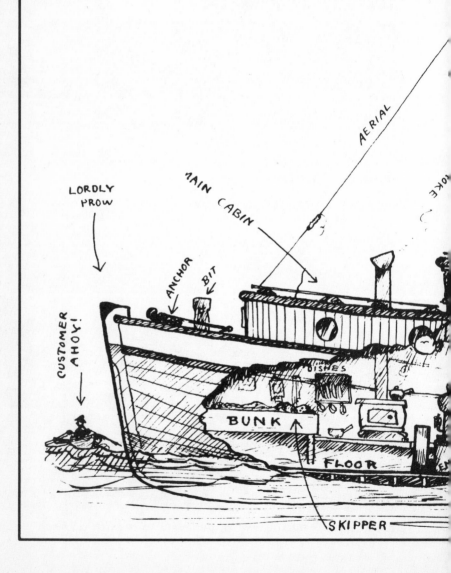

A drawing I made of the Mary in 1935, after renovations.

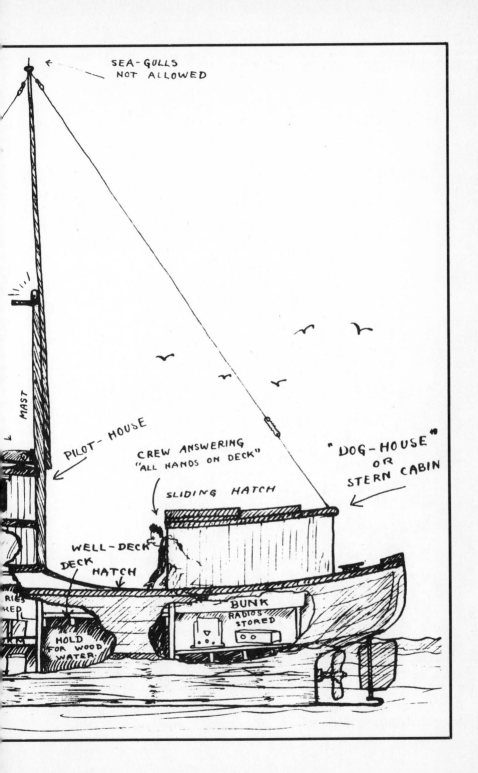

SEA-GULLS
NOT ALLOWED

MAST

PILOT-HOUSE

CREW ANSWERING
"ALL HANDS ON DECK"

"DOG-HOUSE"
OR
STERN CABIN

SLIDING HATCH

WELL-DECK

DECK
HATCH

BUNK
RADIOS
STORED

HOLD
FOR WOOD
WATER

Hob Marlatt, Alan Mace, and I often celebrated the end of summer by going on a boat trip. Here Hob stands on top of Mount Estero.

he'd taken in trade! He had player pianos, he had automobiles, he had refrigerators, and to top it all off, he had a cow tied up—with hay. He just loved trading.

There was a fellow in Powell River named Ross Black. He'd just finished university, where he'd taken mechanical and electrical engineering. He got a job in the mill capping rolls at $3.20 a day. He'd glue paper caps on the ends of the big rolls of newsprint. He'd be flipping these big two-hundred-pound rolls end-for-end all day long, then he'd come and help me in the evenings. He was very interested in electronics. I'd be out doing installations all day and then do shop work with him at night. He'd join us for supper and we'd start working at seven o'clock and work till midnight. I learned an awful lot of the theory that I was missing from Ross, especially concerning the mysteries of AC power. We stayed there until Gordon got into financial difficulties—he was never out of difficulty, really—and couldn't pay us any more, then we stayed on for another month just for the experience. Ross later became mechanical superintendent for the mill.

My spell in Powell River didn't get me any closer to having five hundred dollars a month in my pocket, but I came away armed with some new techniques and, of more immediate use, a completely new understanding of what it took to get out and sell. Up to this time I had been really just sitting back on my haunches waiting for people to hear about me by word of mouth and then find the time to make their way over to Savary and track me down. The sole effort I had made in the direction of sales was nailing up my sign on the Ragged Islands. The area I had been reaching in this way was by and large limited to about a 25-mile radius around Savary Island. Now, coming back to it with my eyes opened by Emil Gordon, and by the urgency I felt to get earning some more money fast, I was struck by the idea my market could be greatly extended if I could somehow get on the water and take the service to the people of the coast. I had to do something, or else pack up and head for Vancouver, and inwardly I was still backing away from that.

I was intrigued by the possibilities right there on the coast. I kept thinking of all the little camps and cannery towns I'd seen on my expeditions with Hob and Allan. I could see this untapped market as far as the mountains went up the coast. Behind every mountain there was somebody with a few bucks in their pocket and they wanted radio. Nobody else knew how to get to them. They could only handle it through the Vancouver department stores. Marshall Wells was selling De Forest Crosley radios, Mac & Mac was selling Westinghouse radios, Eaton's had some line they were pushing. A lot of people would mail in and get a radio, and as I say, nine out of ten were incapable of pulling a signal in up there where signals were so weak. People were at a loss, they were sitting there waiting for me to come and take their money. I felt that I was unique in that I knew how to get around the coast and if I only had a boat I could prove it. But the idea of owning my own boat was like going to the moon then. I had no money to afford it.

It was dear old Frank Osborne, my machinist friend down at Lund, who saved the day. I was fixing a radio I'd built for him, and I guess I was letting out my frustration about not being able to get up the coast, so he said, "Why don't you see old Eric Nelson about his boat?" Nelson, a retired Swedish fisherman, had a fishing boat tied up to the float doing nothing, but needing pumping out every day, and the old guy was getting tired of the chore. He agreed that I could use the boat whenever I needed it, and I would pay him one dollar a day. I must keep it in repair and pumped out.

It was a fairly typical "codfish" boat, built by a Finn in Sointula about 1910, wooden of course, and called the *Mary*. It was 32 feet long with compromise stern, flush deck with a small wheelhouse forward, live-tanks and a removable deck amidships, and a removable "doghouse" in the stern. There was a Little Cod woodstove in the doghouse and two narrow bunks under the foredeck, accessible through the wheelhouse. The wheelhouse contained the engine, which filled it up. The scuppers in the live-tanks had been plugged, so this compartment was reasonably dry and was used to store firewood for the stove.

The engine in the *Mary* was really something to behold. It was a Buffalo nine-horsepower, four-stroke, heavy-duty single-cylinder marine engine with a Joe's reverse gear. It was designed to turn 400 rpm but I usually ran it at 360 to cut down the vibration and conserve fuel. The hull was only capable of 5½ knots at best. The single water-cooled cylinder stood nearly four feet high. The huge flywheel was thirty inches in diameter, and the crankshaft was 2½ inches in diameter. The only adornment on the top of the cylinder was the Champion X spark plug, and down on the starboard side there was a brass priming cock. The Schebler marine carburetor graced the port side of the engine and was not encumbered by any remote controls. The throttle lever would be adjusted for your intended running speed and then left alone. The needle valve, once properly adjusted, was never altered. There was no choke. You used the priming cup, or just put your hand over the intake horn, to get a good suck out of it as you rolled the flywheel. Immediately above the engine, screwed to the cabin top, was a Jefferson spark coil with a vibrator on the end of it. The high-tension wire hung down to where it was attached to the spark plug terminal. A copper knife switch was screwed to the cabin top in a place you could reach it from the steering position so you could make "dead stick" landings. Six #6 dry cells, tied together with fishing line and secured in the bulkhead, completed the assembly.

I received explicit instructions on how to start the engine, and was duly warned as to the consequences if I didn't observe them closely. You used a Johnson bar, which is something like a four-foot crowbar with a large claw on one end. This claw fits around the end of the crankshaft and engages the keyway, so you gain a four-foot purchase in rolling the engine up against compression. If at top dead centre the engine fires in the right direction, the claw releases from the keyway, and you lay it gently aside ready for the next time. The Johnson bar is steel and weighs about 25 pounds. What you have to be extremely careful about is to be sure that you set the timer, the spark advance, just past top dead centre so the engine does not backfire. Eric told me that the previous owner of the *Mary* made this mistake, and when it backfired, the Johnson bar came down across his legs, breaking both, and that's how Eric came to buy the boat for five hundred dollars. I listened attentively and made sure that I always stood with my legs clear of the arc of operation of the Johnson bar.

The Buffalo never failed to start when you did everything right. The procedure was quite simple. First — make sure the clutch is in neutral so it doesn't climb up on the dock when you start. Then turn on the fuel valve. (Don't do this sooner than necessary or the carburetor will overfill and drip gasoline in the bilge.) Then take the squirt can containing raw gasoline and fill the priming cup, open the cup, and roll the flywheel very slowly in the right direction and it will suck the gasoline in in great gobs. If you roll it in the wrong direction, it blows it all out in your face. Only experience will guide you in this matter. Administer about one eggcupful in this way, and then close the priming cup. Now set the timer very carefully in the retard position (just past top dead centre) and close the knife switch (ignition to you). Now take a deep breath, engage the Johnson bar and, bracing your legs against the side of the boat, heave it up and over centre. Just as you do this you will hear a gentle buzz from the spark coil and then kerthump — suck — kerthump — suck — ker-thump — suck.

Not all power boats of that vintage were blessed with having a clutch and reverse gear like *Mary* had. I revelled in this luxury. After getting the engine started, I could take my time and wander out on deck, let my lines off the dock in a very dignified manner, and then ease the clutch in and we were under way, kerthump, kerthump. . .

For those boats without clutches it was anything but dignified. It was downright frightening. You would see the guy let all his lines go, and then with a pike-pole push himself out from between all the other boats till he was in the clear, and then get an oar and paddle around until he was aimed in a safe direction, and then address himself to getting the engine started, only to find that he had forgotten to open the gas valve, or there was a wire off the batteries, and by the time he has all this sorted out, the boat has turned and pointed right at the dock, or has drifted up on the beach. Of course when his engine does fire, he has to make a mad dash up to the wheelhouse and steer like crazy till he is clear of obstructions. Making a landing was even more frightening. The trick was to carefully judge the weight of the boat, the current, and wind, and then pull the switch some distance out from the dock and coast in to a dead-stick landing, very often ending up sitting astraddle the bow with both feet out to ease the shock, and everybody on shore running around and shouting.

Mary not only had a clutch but also a reverse gear, and this was sheer luxury. Mind you, a Joe's gear was only forty percent in reverse. That meant that in cold weather you could go in reverse and rev the old engine up, thump-thump-thump, and the gears would go rear-rear-rear, and the boat would gradually — very gradually — start to make sternway.

In other respects *Mary* left something to be desired, but these things were not really her fault; they were mostly the result of neglect on the part of the owners. I found that in a head sea the whole bow tended to lift up and back, and as it did so the beam increased, because the deck beams had rotted at the ends and were no longer secured to the ribs. This resulted in opening up quite a gap around the bottom edge of the wheelhouse, which let a lot of water in. Of course it closed up again before the next wave came along. By getting inside and lifting with my shoulders, I found that the only thing holding the wheelhouse to the rest of the boat was the rear wall, or bulkhead, which went right down into the fish well and had not yet rotted. This gave me some assurance that I was not likely to lose the wheelhouse entirely, but I decided that if I was to keep using the boat, I would have some proper repairs made as soon as I could afford it.

In the meantime I partly cured the problem by getting some old Dominion Government #8 telegraph wire, tying port and starboard ribs together, and tightening it up with a marlinspike, Spanish windlass style. This stopped them from spreading so badly, but of course it still leaked around the deck line. I hardly noticed this though, because the rest of the cabin leaked when it rained. The canvas deck covering was very rotten. This I found was the main reason she needed pumping so frequently. The hull was reasonably tight. I applied the usual fisherman's cure to the deck leaks. I saved all my Campbell's soup tins and deployed them at strategic locations to catch the drips by the simple procedure of bending the lid up and nailing it to a deck beam. No one even commented on this, as it was common practice with most older boats. The uninitiated, of course, would bump their heads against a brimming tin and get a shirt full.

TEN

I Take to the Water

THE COMING OF *Mary* opened up for me a whole new world of business enterprise, as well as bolstering my ego immeasurably. I invited others along to join me in these trips of discovery, being careful always to choose a companion who didn't mind the cramped quarters, wet blankets, and meagre diet. There wasn't too much you could cook on a Little Cod woodstove. I ventured north through the Yaculta Rapids as far as the village of Roy in Loughborough Inlet, then across through Mayne Passage to Rock Bay on the Vancouver Island side and on down through Seymour Narrows to Quathiaski Cove and back across the gulf, in a kind of circle route. All the way along I would be ducking in and around the islands of the Inside Passage: Cortes, Quadra, Sonora, Maurelle, Stuart, the Redondas, the Thurlows. Every few miles along the coast in those days there would be a small logging camp with up to a dozen men, and quite a number of large camps with steam railroads, employing several hundreds, such as Rock Bay, Menzies Bay, Duncan Bay. Before long I met and knew just about every camp foreman, head logger, fisherman, and stump-rancher in the whole area, and kept an updated record of what radio each person had, when they would need new batteries, and when they last had their tubes tested.

It was so much fun in those days, going around. Every little place like Gorge Harbour on Cortes Island probably had eight or ten families living in it, all totally different. I stopped in Gorge Harbour on one of my first trips. I had talked Hob Marlatt into providing company this time around, and we tied up at a wharf owned by Mrs. Corneille, who was trying to run a lodge there in those days. There was an old renegade Englishman named Kendricks nearby and he had a radio problem. He was one of these remittance men, and his one remaining productive act in life was the making of homebrew. He was always in debt, and it was always on a promise when I fixed his set. One time when he needed a set of B-batteries he traded me a set of leather-covered English binoculars, the kind they use at the horse races.

To reach his place we walked up from the dock through the trees, opened a gate in an English-style fence, and there he resided in a ramshackle sort of cabin with his wife. He had a great big old De Forest 801 and I started pulling it apart. There was nothing seriously wrong with it, and as I worked, Hob sat and chewed the fat with our host. He had one of these loud blustering voices.

"Glass o' beer, chaps?" he rumbled. Neither Hob nor I drank beer, so the old boy came out with tumblers full of a drink he said was raspberry cordial or something, and being as the weather was hot, we took it. In fact we drank it straight off, and before we knew it, the tumblers were full again. This was repeated four or five times and I began to have a hell of a time trying to figure out what I was doing. I was seeing three tubes where I knew there was only one. I didn't know what was the matter with me, but this bug juice he was feeding us must have been overproof, and we were all pie-eyed. Finally I told him I was going to have to take the set down to the boat to try and figure it out at my workbench.

"Okay, boy. That's fine." He didn't even care by this time. He didn't even know he owned a radio. It was a great big chassis, and I had parts everywhere, but somehow we got it bundled up and down the hill. The next morning as we tried to reconstruct our progress we found that we had absolutely no memory of opening or closing the gate. We thought we must have floated right over top of it. Back on the boat, we started preparing supper. Then we had a visit. Mrs. Corneille was a very strict person and her daughter, a nice but very prim girl, rowed out. I forget what all happened, but we apparently made a soup that ended up with radio parts in it — I don't know what she thought of it. Then we decided to go swimming in the dark, and went over and jumped off a diving platform they had anchored out. When we woke up and looked out of the boat the next morning the platform was high and dry on the mudflats. We could have killed ourselves.

When I got a sober look at the set I found I had done the damnedest things to it. It took me the better part of the day getting it straightened out and back up to Kendricks, who of course didn't pay. We were telling about this the next place we stopped and they said, "Oh God, you should have been warned. Any time you go up to Kendricks', look out. He'll get you drunk. Last week the Reverend Greene came in on the *Rendezvous* and they had to get a neighbour with a wheelbarrow to get him back on board his boat!"

English remittance men were one of the hazards of travelling the coast in those days. It was said they were black sheep paid to stay away from England but it seemed there ought to have been some better explanation, as there were so darned many of them. It was hard to feature one small island the size of England generating so many defective offspring, or having so many families wanting to undertake this rather drastic method of preserving appearances. In any case the coast was littered with them during the years between the wars. Duncan was a special centre, but you could run into them all the way north, living like savages in little hovels made of bark but still arrogant as kings.

I had another memorable encounter with a remittance man over on the west side of Read Island. Surge Narrows was a real going concern in those days. It had two stores. I always called, and on this occasion I found a message from a Captain J. Forbes Sutherland (Ret.) requesting I stop and see him. He had a log cabin on Hoskyn Channel, just below Surge, and I duly presented myself. "Oh, how fortunate you called," he said. "I had just been about to purchase a radio on my own, but I'm at a loss to choose the model, there are so many..." I went through the list with him, and he wanted nothing but the best. He had to have shortwave so he could tune in the BBC news broadcasts, and it had to have superior tone, as he had, by his own admission, a highly cultivated ear for music. He also made reference to the exquisite taste of Mrs. J. Forbes Sutherland, whom he had married since coming out to Canada. She was a Trory, of the well-known family of Vancouver jewelers, but he said she was temporarily away. I ended up selling him a Stromberg-Carlson, which was my top-of-the-line set. I went up a tree and rigged his aerial, got him all set up, hid all the wiring so as not to offend the delicate sensibilities of his wife, and he had me in for a truly delicious curry dinner which he had learned to cook in Injuh. I left feeling the English were after all a noble race.

A week later his cheque bounced.

This was a matter of about two hundred dollars, more than all the profit from a whole trip. More than several trips. I couldn't afford it. I went straight back, planning to retrieve the set while it was still new. "Oh, what a nuisance," he laughed. "These fool bankers in this country. . ." He assured me there had been a silly mix-up between his account and his wife's, and all I had to do was go back and present the cheque again. I fell for it, and of course found no one would touch his cheque. I now discovered he was known up and down the coast for pulling this sort of thing. He had been living off his wife, but she'd had enough and pulled out. He had long since exhausted his credit in both Surge Narrows stores and had begun rowing further and further afield, blustering his way into debt with storekeepers at Heriot Bay, Campbell River, and Refuge Cove. He'd even rowed as far as the Wilcock's store at Stuart Island.

I couldn't afford to give up and went back, determined to get something. I dropped anchor, launched the dinghy and rowed ashore. When I reached shore, he called to me. He was standing at the head of his long set of steps with a double-barrelled shotgun.

"You do not set foot on my property, Spilsbury," he growled. The gun's old English-style hammer made a sharp snap as he cocked it. I was three miles from the nearest help. I retreated and in great distress went to my friend Roy Allen of the Provincial Police in Powell River.

"What do I do about this?"

"God, you should never have got tangled up with that guy. Didn't anybody tell you?"

Roy sent me to Tommy Taylor, who was the resident lawyer there. But Tommy said I needed a Vancouver lawyer, and he happened to have the name of an eager young fellow just starting practice, Esmond (Bud) Lando. On my next trip down to Vancouver I went up to the Hall Building, introduced myself to Bud and told my sad tale. Bud explained that it didn't matter how good my claim was if the guy didn't have any money and advised me to get in touch if I ever heard of Sutherland getting back in the chips. About a year later one of the Jones boys took a few sections of logs off Sutherland's property, and as soon as the boom looked ready to go, Bud plastered it. Sutherland ended up paying me in full, but more than that he had unwittingly put me in touch with a person who was to play an important role in all my future dealings — Bud Lando.

Down the channel a little ways from Sutherland was Francis Dickie, the writer. Dickie wasn't a displaced Englishman, although he probably would have liked to have been. He was born in Manitoba, about 1890 I should think, because he was in his middle age when I first knew him. He thought a lot of himself. He'd been in Paris at the same time as Hemingway and never got over it. He was very proud of having the odd letter from Somerset Maugham and brought it up often in conversation. He came to Quadra Island determined to write great things, and spent a lot of time strolling around his garden with no clothes on like William Blake. He had a tin-can bell with a rope running out to the gate that you were always supposed to ring so he could dash in and put his pants on. People tried to be good to him but he took all favours as his due, and eventually put most people off. One of his books was called the *The Master Breed*, and you would sometimes hear people referring to Dickie himself by the nickname "Master Breed." His writing was dreadful, I thought. His wife Suzanne was a Parisian woman who was all the things Parisian women are supposed to be and was often the subject of unkind speculation amongst the neighbourhood stumpranchers. Many years later a woman from the area told me Suzanne had been lusting after me, in a very distasteful way apparently. I have to take their word for it, because it wasn't distasteful enough for me to notice.

Read Island had settlements on two sides, with Surge Narrows on the west side and Burdwood Bay and Evans Bay on the east. There were no roads, so there was little communication between the sides. The Read Island post office was in the middle of the island and equally inaccessible to all. From the steamer landing on the west side you had to walk an old trap line, cross a few farmer's fields by climbing over barbed wire fences and after two miles there you were at the home/office/barn of John Jones, an old Welshman who had bought about six hundred acres at the turn of the century. He had been postmaster for as long as anyone could remember.

I met him on one of my first trips in the *Mary*. I tied up behind the government steamer float that was anchored in the bay in front of Bill Frost's store at Surge Narrows. I had letters to post and asked Bill Frost if there was a way to do so without making the excursion to mid-island. He told me I could give the letters to John Jones when he came down to meet the weekly steamer. The boat was due in with freight and mail sometime between midnight and daylight the next morning and Jones never failed to meet it. I had some freight to ship so I made out the waybills and took it to the shed. The door of the shed was closed, and when I opened it, the inside was quiet and dark. I went in with a flashlight and looked around for a place to deposit my load. There was a lot of junk lying around — empty cans, fuel drums, a few coils of wire rope, and in one corner what looked like a large pile of gunny sacks.

It was too cold to stand around and I was about to leave when I heard a cough and the pile of sacks moved. As I watched with mouth agape, poised to bolt out the door, the sacks continued to shift and slowly unwind until John Jones was standing before me. He was there waiting for the steamer as usual, but he believed in comfort. He had been sitting on a crate with his old coal oil lantern between his knees and the sacking pulled up over him like a tent. A humid cloud of kerosene fumes and body odour escaped past me when he opened up.

I stated my needs, he fumbled mutely in the bag draped over his shoulder, brought out stamps, and postmarked my letters. Then with hardly a word spoken, he lowered himself back onto his crate and disappeared beneath the sacking. He was not the yacky type.

When the steamer finally pulled in at 5 A.M., John Jones stuffed the incoming mail into his bag, rowed ashore, and hustled up the trail so he would have the mail all stamped and sorted before his first customers arrived.

In those days the Union Steamships started using the radio to advise their upcoast customers of estimated times of arrival at all the different stops. This was a godsend to people in places like Surge Narrows, because all they had to do was put their radio dial over to 1650 at the specified time and they would know whether their boat was coming at ten in the evening or six the next morning. For people who had radio, it was no longer necessary to wait around all night in cold weather the way John Jones did. For John Jones, however, it made no difference, because he had no radio. Several people had tried to sell him one, and he was reputed to have a lot of money stashed away, but those who knew him said he was too tight-fisted to spend it on a radio.

Somehow the old guy got wind of this gossip and it got his bristles up. He left word with Bill Frost that the next time I came around he wanted to see me about one of these new-fangled gadgets everybody was yapping about. I told Bill that if the steamer broadcasts were the only thing he wanted, I could build him a little set for twenty-five dollars.

"Hell no," Bill said. "Let him pay. He's got lots of money."

Next time I was by he sent word that he was coming down to see me. The whole community was talking about it. People were taking bets on the outcome. When he appeared, I didn't know quite what to expect, so I started by showing him the smallest and most modestly priced set I had on board. I could see it didn't interest him, so I brought out a couple of trade-ins that were even less. I assured him these would be more than adequate for monitoring the boat channel but I don't think he even listened. I fell silent. I wondered if he would take one if I gave it to him.

"Never mind all that," he said impatiently. "What is the most expensive set you have with you?"

As it happened, I had a very expensive set with me. It was an eight-tube RCA Victor with all bands, standard and short wave, in a luxurious console cabinet the size of a small refrigerator. I had splurged on it not knowing if I could sell it, and so far no one had even wanted to look at it. It was still in the carton. John Jones didn't want to look at it, either.

"Just bring it out to my place and install it. If it works, I'll pay cash," he said. Then he got in his dinghy and rowed away.

Now I had a problem. His place was two miles away by goat path. The set was much too heavy for one man, and very delicate. Once I got it there, it might not work. You never knew until you tried if a given set would work in a given location.

Bob Tipton, the other Surge Narrows storekeeper, came to my assistance. The Tiptons had a horse, and we rigged an Indian-style litter consisting of two poles tied to the horse's collar with butt ends dragging. We tied on the bulky carton and Bob detailed his stepson Reg Keeling to give me a hand with the horse wrangling, which he could see was not one of my specialties. I just hoped the set would stand the shaking around. By taking down a few fences we were able to get right to the Jones house.

People told me I would know the place when I saw it, and they were right. It was set in the middle of an outcropping of bare rock and it looked like a bomb shelter. Apparently his original farm building had burned a few years before and he was determined this wouldn't happen again. After blasting the bedrock flat he poured a floor and half-walls of concrete. The upper walls and roof were corrugated iron. I don't recall any internal partitions. There seemed to be just one large room with a stove and kitchen table in one corner and bedding on the floor in another where it appeared he and his wife slept. Another corner was reserved for His Majesty's Service, with postal scales and a stock of empty HMS mailbags on the floor. Opposite it were sacks of grain and meal for the goats and chickens. There may have been a chair, but I wouldn't swear to it.

Jones grunted recognition and indicated a place along the wall, recently made bare, where the set was to go. Mrs. Jones was hovering about but he never once spoke to her or made the faintest sign of recognizing she was there. She smiled and nodded constantly, making pleasant noises of approval, but spoke not a word. I gathered the poor soul was a bit balmy.

It took only a couple hours for me to unpack the set, erect an aerial, hook up the batteries, and show him how to operate it on both standard and short wave. Among the short wave stations, BBC London was coming in like a local. We were miles away from the nearest source of electrical interference, and domestic stations were as clear as you ever heard them. The set's big speaker set that cavernous room ringing like a concert hall. Just that simply, the remote interior of Read Island had been made a part of the modern world. There were times like this when the wonder of radio science struck me anew, and made me swell with pride in my trade.

"How do I get the steamer?" John Jones asked.

I had completely forgotten about this incidental matter. I decided to do it up properly and used my portable signal generator to set the dial precisely at 1650 kHz and marked it on the glass with waterproof ink. Fortunately, before I left we heard one of the steamers down the coast announce its arrival times.

"That will do," he said, "Leave it set right there. Just show me how to turn the thing on and off. I don't need any of that other foolishness."

This was all he wanted the set for. Neighbours told me months later that the dial was still set at 1650. The only reason he'd insisted on the big RCA was to prove his critics wrong for calling him a cheapskate. It cost him $220—no small change in the dark days of the depression. I carefully added up the invoice and he just as carefully checked it over. Then he went over to His Majesty's corner and produced an array of tin biscuit boxes and tobacco tins. They were heavy. Inside were coins—a few fifty cent pieces, more quarters, a great many nickels and dimes, and thousands of pennies. This is how I was paid. He would count them off one at a time into a pile which he insisted I recount and approve by initialing a little chit. It took longer to count the money than it took to install the set. I was glad I had the horse to help me get it back to Surge Narrows. He had been taking in people's change for stamps all those years and when he saw me he thought here was his chance to finally unload it all. When I complained about the cumbersome method of payment back at the wharf, Bob Tipton and Bill Frost laughed and said I didn't know how lucky I was — it was the first time they'd known any salesman to succeed in selling John Jones *anything*.

A lot of the men who lived on the Read, Stuart, and Cortez Islands in those days were returned soldiers from the Great War. They weren't yet referred to as veterans and wouldn't have appreciated the implication of age that term implies. They were in their prime.

Being in their prime, they would periodically get together to celebrate the fact, and this would result in a wild party, with any luck. During the year they were noted for a fairly constant level of bickering and squabbling and feuding amongst themselves, but when the party date came they were comrades in arms once again.

I set off on my rounds one time only to find out that the "boys" were off on a big wing-ding at Whaletown. Many of my regular customers — Squirrel Cove storekeeper George Ewart, Bert Wilcock of Stuart Island, George Byers and Baldy Martell from Blind Channel, Bill Frost and Charles Redford at Surge Narrows — were included. One man who regularly promoted and conscientiously attended these affairs was Bob Tipton of Surge Narrows. Tipton was a justice of the peace, and both he and his wife Nell were fine specimens of the English upper crust, but Bob would let himself go on these occasions and come back to Surge Narrows very much the worse for wear. This landed him in deep trouble with Nell, who was not a returned soldier and did not let herself go. Finally she laid down the law and said, "No more!" Bob was ordered to stay in the house under 24-hour guard while Nell sent word to the organizers that her husband no longer had any interest in being a part of their tomfoolery.

The gang was not about to take her word for it, and in due course a boat with ten or fifteen of them crowded aboard pulled in at Tiptons' dock and blew its whistle. Mrs. Tipton appeared on the front porch in a billowing nightgown, waving a corn broom and shouting at the gang to get going. Bob was not coming. Unlike them, he had some slight sense of responsibility to his family and his business.

"Aw, come on, Mrs. Tipton," they chorused.

"Bob doesn't want to go," she insisted.

"Let him tell us, then!"

Just then Bob appeared around the back of the house in his underwear, a bundle of clothes under his arm, and broke for the float at full gallop. She saw him and bolted after, the broom raised menacingly. He had a few steps on her, but she was putting more ferocity into it. The gang urged him on at the top of their lungs. Then, just as she was drawing within range she struck wet grass and her feet shot out from under her. The nightgown spread out like a parachute, revealing all it was meant to conceal, and she landed hard on her back. Sensing his own health to be more at risk than hers, Bob kept up maximum rpm and made a flying leap into the cockpit as the skipper slammed in the clutch and churned white water astern.

I made many friends and very few enemies. After all, how else could they get their precious radios fixed? I had virtually no competition except from the wholesale supply houses such as Marshall Wells and Mac & Mac, who would sell direct to the logger at wholesale prices. It was very usual to find someone in trouble who would exhaust every possibility before my arrival—send to town for new batteries, then a complete set of tubes, and when this didn't make it work, they would mail one of my self-addressed cards requesting me to call on my next rounds. At this time they would wheel out all this new merchandise with great pride, not realizing that this represented lost revenue to me. The average cost of a set of dry batteries was $25 to $30, and a set of tubes anywhere from $25 up, on which my discount would have amounted to 25 percent, or probably $10 to $12. Two such sales a week would keep me in business. As it was, I had to establish very firm rates for my services, and that raised a few eyebrows—75 cents an hour, and not one cent less! Depending on the complexity, most jobs, such as replacing an audio transformer, retune and cleanup, would run two to three hours, including a clean and polish job on the cabinet—two to three hours plus the cost of the new transformer, so the total bill would be about $7.50. If I could average two such jobs a day, I was in gravy. This took care of all travelling expenses, and the renting of *Mary*. Incidentally, at that time I did work in the customer's house whenever possible, having little room to work on the boat. This provided them with several hours of free entertainment, which generally involved all the neighbours, but if I was lucky I got invited to eat with them and this was a real bonus.

One of the smartest sales gimmicks I had was cleaning and polishing the cabinet when I got through. Most of the sets in those days had very large, ornate walnut cabinets and I usually found them looking extremely grubby, covered with fingerprints and cooking smoke. I would leave them gleaming like new. This always went over very big with the lady of the house and she would beg me for the secret. I used, very simply, a lemon-oil polish supplied by Williams Piano House. It had no label and was quite inexpensive, and I finally got smart and ordered the stuff by the case, put my own tube-sticker label on it, and obligingly sold this to the appreciative housewife at 75 cents a bottle. This became a major source of revenue for me. Sometimes when I was called in to just do a routine checkover, for which I charged one hour, and couldn't find anything actually wrong with the radio, I would give it a polish job and the owners would, without exception, agree that the radio sounded twice as good. I would add 75 cents to the service charge for a bottle of lemon oil, and go back to the boat with $1.50 in my pocket.

I didn't have any bookkeeping. All I knew was the money that went out and the money that came in. I didn't analyze it or classify it. Part of the money that went out, when I was on the boat, had to buy food for the trip. I kept track of every can of beans and every dollar I spent in the store, and the money I had to buy gasoline with, plus the dollar a day for the boat, plus the cost of the equipment that I had bought—and those were my total expenses, as I understood it. I put that against the money I collected, and on one typical three-week trip I calculated I was clearing $1.54 a day. This I considered net profit, an exciting idea. I could have been working wheelbarrowing gravel on Savary for $3.20 a day, but that wasn't profit. I considered myself hugely successful, until I tried to calculate how much more I needed to reach the magic five hundred dollars per month.

I'd see the girl a lot every summer and her parents, I think, started to get quite nervous about her interest in this rough logger fellow up the coast. If only they had known, they would have left well enough alone. Our involvement was quite platonic. No hanky panky whatever. I didn't know enough for that. Nowadays it is hard to believe just how innocent it was possible for young people to be as recently as 1935. On one of my trips in the *Mary* I had an experience which illustrates this. I was calling at Thurston Bay on the west side of Sonora Island, where between 1914 and 1941 the BC Forest Branch had their main shipyard and maintenance depot and I always

The World War I vets of Read, Cortes and Quadra on the way to their yearly wing-ding.

Looking up Phillips Arm from Shoal Bay, 1932.

Storekeeper Sid Vicary at Deceit Bay, 1932. Former Fraser River paddle-wheeler Transfer *supplied village both accommodation and steam.*

BC Forest Branch maintenance depot at Thurston Bay, 1932.

found some work. As I walked past one of the cottages, I glanced up at the window and there, sitting on her bed with no curtains drawn, was one of the workmen's wives, nude from the waist up. I quickly averted my eyes, but the brief glimpse threw me for a loop. I had never seen a woman's breasts before. Women in those days somehow managed to give a convincing outward impression their bosoms were little different from a man's, and I'd heard nothing to make me doubt that this was more or less the case. I left Thurston Bay in shock.

And this was when I was nearly thirty!

The girl's virtue couldn't have been more secure than it was in my company. Which was fine with her. She was just as innocent as me at the time. But her family thought they knew better and they shipped her off to a convent school in Switzerland for a year. The day she returned to Savary, I can remember it so well. She arrived in an airplane. One of those old, old seaplanes, probably a Fairchild 71 or something, and it made a big impression on everybody. She had a new boyfriend, some guy with a lot of money to spend on her, and he had chartered this plane in Vancouver just to bring her up. We were still ostensibly interested in each other, and I went out to see her but I said, "I can't talk to you because I've got whooping cough, and it's very contagious!" Not much of a way to make an impression, but that didn't stop her. She didn't give a damn about this other guy, she just got him to fly her up there and gave him the brushoff immediately she arrived, _for me_. And had she ever changed! She knew all the tricks. I was afraid she was going to seduce me on the spot. That's what convent school did for her. She stayed with friends for a while, but I was scared stiff the whole time. I didn't know what to make of her. So she went back to town and first thing I knew she was married. But in the years to follow, between husbands, she'd come and take another look at me.

I felt lucky to get off as lightly as I did. But boy, for a few years there she just knocked me endways! I never wanted anything in my life as bad as I thought I wanted to get this five hundred a month so I would be recognized by her parents.

It had its overall effect on my life and I think it probably speeded things up considerably. The five hundred dollars was so completely unattainable, it was an absurd figure. Whether her uncle actually told her that, I wonder. I think she just wanted to fire me up. Get me out of there. And bigosh, she succeeded. It sparked something in me so I went out into the world and said to hell with everything, I'm gonna do it! I think my parents were as worried as hers then. Dad didn't say very much but he didn't really approve of the kind of company I was showing an interest in.

ELEVEN

The Radio Boat

I CARRIED ON with Eric Nelson's old *Mary* for several years, but found increasing difficulty trying to keep my merchandise dry under the leaky decks, and also having no room on board to do service work. I had approached Eric Nelson about the possibility of his fixing it up, but he was not prepared to spend a nickel on it. He said I could go ahead and do it myself if I wanted to, though. I decided to have a talk to Hugo Johnson, the old Norwegian boat builder. Hugo had been building and repairing boats at his little shipyard in Lund for years and was considered a master of the craft. He built some distinguished ships including the Thulin brothers' tug *Niluht*. I remember being called in to witness the christening of the *Niluht* with my dad in 1922, when old man Thulin explained how they came up with that name.

"For a long time we figured *Flor de Liss* would be a nice name," he told the gathering. "Then we got the idea to put Thulin backwards, and that sounded nicer." The Thulins were the only ones who thought so, and when they sold it the first thing the new owners did was change the name to *Shepody* after Shepody Bay. The boat itself was a complete success, for it withstood half a century of beating up and down the coast and was still doing good service in 1987 under the name *Viking King*.

Tuning up the *Mary* was a breeze for Hugo, so we put her up on the ways at Lund and in the amazingly short time of about ten days, Hugo finished the job at a total cost of $450, including labour and material. When he got through, no one would recognize the old *Mary*.

First, he ripped off the cabin and the decks from stem to stern, so only a shell was left with the big old Buffalo sticking up in the middle of everything. He replaced the oak bowstem with a new one about eighteen inches higher, then he doubled up all the oak ribs, raising them well above the old deck line. He replanked the sides to result in a semi "raised deck" hull from the bow to about two-thirds of the way back. He replaced the main deck with a waterproof one, which gave me a large dry storage area where the old fish well had been. The after doghouse was made permanent, and I fitted it up as a workshop. The new wheelhouse was wider, with good windows. This enormously improved the headroom forward, so now the living space was quite comfortable. I converted the old woodstove to burn stove oil and installed it in the main cabin, along with a small sink and freshwater tank. There was still no head, of course, but there was more headroom to use the overside bucket. To top it off, I acquired and installed a real steering wheel. In the original *Mary*, steering was accomplished by a large vertical wooden lever that stuck up through the floor of the wheelhouse. The hull was recaulked and I repainted the entire ship. I could hardly wait to get back on the run.

Well, the new *Mary* was a revelation in comfort and convenience. Dry blankets to sleep in, no drip cans to empty, and the oil stove...! Jack Parrish, the machinist at Stuart Island, helped me rig it up. Actually very simple when you're shown how. A four-gallon tank of stove oil on the cabin roof, with a quarter-inch copper tube running down to the stove with a simple needle valve, and the end of the tube projecting into the firebox of the Little Cod cast-iron stove; two pieces of wet fir bark (later replaced with a few blocks of lava rock), crumple up a bit of paper, turn the oil on to drip, and light. It ran 24 hours a day and kept the cabin warm and dry. The tank full would last a week, and it was 12½ cents a gallon in those days. It did smoke and build up a lot of soot. In about three days of operation the little stovepipe would become plugged and the stove would not draw. All you had to do was give the pipe several hard whacks with a wrench handle and volumes of this fluffy black stuff would billow out of the stack and float across the bay, falling quietly like black snow over all the boats on the downwind side. If you wanted to stay on good terms with your neighbours you had to choose the right moment, or do it after dark.

I didn't have long to revel in my new state of opulence. Old Eric

Nelson came down to look at the job, quietly announced that it was worth more money now and insisted I begin paying him five dollars a day. He made it very clear that the matter was not open for discussion. I couldn't believe it. I talked it over with Hugo Johnson, who just shrugged his shoulders. He'd got his money. I then confided my problems to Frank Osborne, who had introduced me to Nelson. He was sympathetic, but in no way surprised. His only advice — "Next time you do business with a goddamn Swede, get it in writing." The advice stuck with me over the years — Swedes or not.

In spite of the five hundred percent increase in tariff I stuck it out for a few trips, as business was picking up. Then Eric came down one day and said he wanted his boat back. Right then. He told me he had decided to go back fishing, but I think he really had a sale for it. I was in no position to get in the bidding, so I handed it over to him — and then began a frantic search for another boat. I had customers waiting and expecting me, and there was word getting around of another seagoing radio man named Richardson who was about to start up in competition with me. Most of my customers were very loyal and said they would wait till I could get going again. But what really saved me was that my rival was a little more diversified than I, and included beer and hard liquor in his line of merchandise. He made some inroads, but was chased out of most camps, whose owners didn't want liquor on the premises.

I don't know whether I mentioned it before, but I had earlier become interested in being a transmitting ham and got my licence in 1926. I hadn't seriously tried to put a ham set on the air until I had the licence — I was afraid to — but I didn't waste any time once I did. The term *ham* refers to a non-professional operator whose hobby is radio and who holds an Amateur Experimental Operator's Licence. Originally we were permitted only to transmit dots and dashes and you had to be able to send and receive ten words a minute to get the licence. Later, voice transmission became possible, but it wasn't until much later I installed it.

Different people become hams for different reasons. Some want only to design and build their own equipment, and have no interest in actually operating a station. Some prefer to buy ready-made equipment and get their kicks entirely from communicating. You could talk to different people every day, and build up an association of friends all over the country, or even around the world. Some hams become hooked on working DX (long distance), calling the farthest corners of the world. There are competitions for the most countries worked in a given day or week or month. Then there are those who prefer to engage in handling third-party messages, just like Western Union but for free. Many join regular networks specifically to "handle traffic" and pride themselves on how many messages of this nature they handle per month.

Because of my isolation, my greatest interest was communication. Ham radio got me talking to people off-island, people on the mainland, people in Vancouver. By 1927 I had talked to hundreds of hams including 5HP, Jim Hepburn, down at Victoria, and all the hams in Vancouver including Gordon Armstrong, 5UP, who had one of the oldest licences in the group and who later sold radios for me. I got to know a hell of a lot of these people just meeting them over the air and talking to them every night. We'd all get together and rag-chew: "Where's so-and-so tonight? Oh here he is, he just came in from Bella Bella..." and so on. All in Morse code of course. It was quite a fraternity.

I soon found myself handling a lot of third-party messages. The Dominion Government installed an underwater telephone cable to Savary fairly early on, but it would periodically get broken and the land lines on the island, which were strung between the trees, often went down in storms. Then I would get on the air, call "any station in Vancouver," connect with a ham there, who would get on the telephone and close the gap.

Radio amateurs the world over have distinguished themselves by providing essential services during times of emergency when the normal lines of communication are down. Even today. They gave first reports of the Alaska earthquake of 1964 and reassured relatives after the Edmonton tornado of 1987. I loved it. The feeling of importance that went with it, and the sincere expressions of appreciation from those you helped.

In the winter of 1935 western Canada was hit by one of the worst ice-storms in history. It was called the Great Silver Thaw. Prolonged zero temperatures were followed by a sudden change to rain while the ground was still frozen. The rain froze instantly to everything it touched—roofs, trees, roads. Power lines and telephone lines collapsed under the weight of ice throughout BC. It took weeks to restore full service. For three days virtually the only means of communication between the west coast and the rest of Canada was ham radio. Two cross-Canada passenger trains were lost without a word. Even the company couldn't locate them, and waiting relatives

The Mary *after Hugo Johnson's $450 rebuild.*

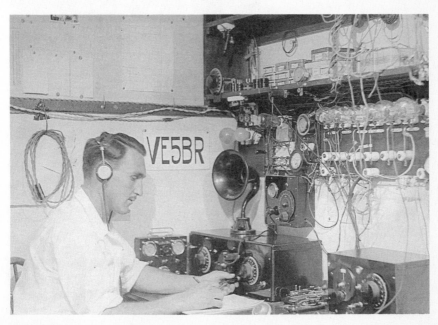

Plugging into my ham receiver in 1931. The gear to my right is test equipment.

My bedroom broadcasting station with ham transmitter in 1931.

had no idea if the passengers were dead or alive.

Within hours the hams were on the job and the messages were flying. One of the westbound trains turned out to be derailed a few miles out of Vernon. The train got word to the CPR agent in town, but that was as far as it went. He had no way of sending word on. There was a ham in Vernon, VE5KN, whom I talked to almost every night. He tried to reach Vancouver with word of the train, but due to electrical noise, hams in Vancouver could not hear VE5KN's weak signal. I could copy him though, so he sent me the message from the lost train, addressed to CPR General Office, Vancouver. I got on the air to Vancouver, but they couldn't hear my signal either. But another ham in Parksville, VE5BL, could hear me, and he was getting through to VE5AS in Vancouver, so the message went 5KN-5BR-5BL-5AS. The reply was quick in coming, and we retraced the steps back to Vernon. From then on the messages flowed for three days, just as fast as we could take them: instructions to the train crew, word to the repair gang, lists of passengers, messages postponing weddings, even some postponing the funeral for a corpse en route to Vancouver. I handled 340 messages in the three days. I went forty hours without sleep, keeping the circuits open with my little home-made transmitter on Savary Island. That's the way ham radio works.

Along with the rest of the hams involved I received a most appreciative letter from C.A. Cotterell, assistant general manager of the British Columbia District of the CPR. "We were indeed pleased with the manner in which you and your fellow operators assisted us in obtaining contact with our various offices in the interior of the province," he wrote. "The manner in which the numerous stations stood by to receive these messages, and the high degree of accuracy in transmission...indicates a useful future for your organization." I pinned the letter over my transmitter. It would prove a valuable asset in time.

Roy Allen, 5MV, the BC Provincial Police officer from Powell River, was a ham in his spare time, which he didn't have much of. Through him I became involved in helping the BC Police network relay messages between their boats up and down the coast, which had mostly homemade telegraph transmitters. Small, key-operated things. They didn't have telephones. They felt too many people would listen in. I operated on BC Police frequency from my station at Savary, which I had a letter from the BC Police in Victoria permitting me to do. I became such a fixture in their system it became difficult for me to leave when I decided to go up the coast, and this is where Mother came into it. She sat down and learned the Morse code in a matter of six weeks, and operated the station all the time I was away. She was sixty or better then, but she mastered the whole process without a hitch. When she decided she wanted to do something, she was hard to stop. From then on, she did all the police work.

The ham style was very important in the way I approached radio, because the ham got every last ounce of action out of everything he bought. He'd make one tube do what a professional would take twenty tubes to do because he didn't have the money to do otherwise. This approach also proved ideal for using radio up the coast because there was no power. With three tubes, using a reflex circuit and adding regeneration, I had a set that they could afford to run—it didn't take much in the way of batteries. Some of these sets they were buying had eight tubes, migosh, they'd drain a twenty-dollar battery in one night. They had to take a storage battery in to the nearest town to get it charged or take it down on the gasboat and run the engine all day. The Parksville ham who had relayed my CPR messages into Vancouver, R.C. (Bob) Weld, 5BL, was one with whom I had frequent contacts. Aside from acting as local magistrate and playing with ham radio, Bob was a boatbuilding enthusiast, having built several small sailing vessels over the years. In the previous year, with his son Brian's help, he had built quite a large powerboat, the *Wyvern III*, in his back yard, launched it in French Creek, and had made one trip as far as Savary Island to show it off to me. Compared to anything I had had, it was a veritable palace, with a head, five bunks, a dining saloon and owner's stateroom. It just made me drool.

After getting Eric Nelson's ultimatum I of course told Bob all about the loss of the *Mary* over ham radio, and at the same time asked if he knew of anything in the way of a boat for sale or rent. When I received the answer I thought I must have made a mistake in deciphering his Morse code and asked him to repeat. He tapped out, "Why don't you buy my boat?" He went on to explain that, now that his son Brian was married and living in Victoria, the new boat was too big for Bob to manage by himself, and he had decided to sell it. He was asking $2,500. When I explained my cash position, he said terms could be arranged to suit. "Like what?" I enquired. He said, "Well, you were paying Eric Nelson a dollar a day just for rent and could manage that, so you pay me a dollar a day for 2,500 days and it's yours." I was so elated I could have flown over to Parksville without an airplane, but as it was, I went down to Powell River in the morning, across on the old *Princess Mary* and down to Parksville by bus. What he had in the way of legal papers we signed and the next day I arrived home with my new boat, which I re-registered under the name of *Five B.R.*, derived from my ham radio call, VE5BR. There was still a lot of finishing work to do inside, so I took the opportunity of getting Hugo Johnson on the job and carried out considerable remodeling to come up with a very practical yet comfortable design that was to serve as my office, workshop, and home. That was 1936. For the next six years I never slept ashore.

With the new workspace, I improved my array of test equipment and it got so it didn't matter what kind of a set came in, I could fix it. I got a little Johnson Chore-Horse generator and this gave me sixty-cycle AC. With this I could operate a tube tester, a signal generator, and a cathode-ray oscilloscope. Another ham acquaintance, Bruce Lanskail, was the chief technician for RCA Victor and one of the foremost radio men in the west, and he sold me the parts for this oscilloscope. It had a one-inch screen. With it I could analyze a set. I could tune it to flat-top the response curves. It was later used by many other technicians and for years you couldn't beat it. As long as I had that little light plant popping, I could do anything anybody could do in a well-equipped shop in Vancouver.

The *Five B.R.* became a fixture in the little camps, canneries, steamer stops and stumpranches from Pender Harbour to Seymour Inlet, although few people ever referred to it by name. There was something about the name being numbers and letters mixed together that people couldn't cope with. Because it also caused confusion with my ham call when I was giving it over the air, I often had occasion to think better of it. Most people simply called me the Radio Boat. I installed a police siren on it which I would blow coming into harbour and everybody ashore would say, "There's the Radio Boat."

TWELVE

The Cannon that Flew over Lund

THERE'S A STORY BEHIND THAT SIREN which involves our old friend Frank Osborne. One time I came into Lund after I had been absent for a few weeks, and the whole appearance of Frank's shop had changed. There was a big new painted sign over the door proclaiming F.P. Osborne to be the *Authorized Agent for DEUTZ Diesel Engines*. More signs were posted around the shop. In front of the office, just out of its crate, was a shining new sample of the Deutz marine diesel engine. Some salesman had done a real job on Frank, and all he could talk about was this marvelous new "High Speed, Two Cycle Diesel" that he was offering.

But that was not all. There had been a much more drastic change in the shop which I found very upsetting. He had done away with his old five-horsepower vertical Fairbanks-Morse distillate engine with the make-and-break ignition that had chuffed faithfully away for so many years, turning all the wheels and pulleys and slapping leather belts in the machine shop. In its place was a new, green, German diesel which was making a piercing scream as it supplied power for the shop. I asked Frank about it. He took me outside so we could get away from the noise and launched into an enthusiastic lecture on the wonders of these new machines. This particular one was quite innovative. It had two horizontally opposed cylinders on a single crank, and turned up at quite high revs. Frank said it would run all day on a beer bottle full of diesel fuel, there was nothing to wear out, and it would run for ever—everyone should have one. All this effusiveness was quite out of character for him, and I can only surmise what kind of salesman had administered the treatment. The old shop didn't seem quite the same after that.

It was a few months later before I came back to Lund, and to my surprise the old Fairbanks-Morse was back in place and going chuff chuff—ha ha ha—chuff—ha ha—just like it used to. There was a large hole in the back wall of the shop near where the Deutz had been sitting. Frank wasn't saying much, but after pressing him for an explanation, he told me the story. The Deutz ran just like the man said for several weeks. Then one lunch hour Frank had gone up to his house, directly behind the shop, when he heard the engine making an unholy noise and getting louder and louder. He was on his way down the hill about as fast as he ever moved, when the Deutz met him coming up.

It had jumped right off its base and gone through the cedar board wall, and "like to chased me back up the hill," as Frank put it. There was a very plausible technical explanation for its behaviour. Being a so-called two-cycle it sucked its air through the base of the engine. For some reason a quantity of lubricating oil and unburned diesel fuel had accumulated in the base over a period of time. This accumulation got sucked in along with the air and ignited in the cylinder and it ran faster and faster, since the governor no longer had control of the fuel supply. As the speed picked up, so did the amount of surplus fuel increase till the flywheels burst and the engine practically disintegrated, but not until it had succeeded in winding up all the old leather belts and pulleys in the shop, leaving the place in a complete shambles. Needless to say, all the signs had come down too, and the sample engine was gone from the floor.

The subject was not discussed again. Frank was not given to that kind of mistake and he did not take it lightly.

I haven't forgotten that this is supposed to be a story about a siren and I'm going to get to that now. On one of my next trips up the coast I called in at the BC Forestry Station at Thurston Bay. The Forestry Branch operated a large shipyard and machine shop here, and did almost all the maintenance and repair work on the fleet of forestry launches. Bill Bachelor was the foreman in charge in those days. I was wandering around in the shop, admiring all the racks of spare engine parts, when I saw a small object that got my curiosity going. It turned out to be a police siren off a Harley-Davidson motorcycle. It had a small grooved pulley on the end of the shaft, which would be engaged with the side of the front tire on the bike when in use. Bill noticed my interest and picked it up to demonstrate it by holding the pulley against the leather belt on the hydro generator. It instantly wound up to a piercing scream, and I decided I needed it badly for my business. Having no real use for it, Bill gave it to me. My plan was make a bracket so I could attach it to the roof of the wheelhouse.

It was for the purpose of getting some machine work done on the bracket that I took it in to Frank Osborne's shop. As usual he was working at the lathe, the old Fairbanks-Morse was going chuff chuff—ha ha ha, and as usual Frank didn't look up from his work. I had the police siren in my hand, and I had a bright idea how to get his attention. I just reached over and held the friction wheel against the drive belt of the old Fairbanks and it produced a most ungodly yowl that could be heard all over Lund. The effect on Frank was spectacular. He jumped off his stool and bolted straight for the Fairbanks. I was standing right in his path and actually held the siren up in front of his face so he could see what it was. It was still yowling. He knocked me over in the process of getting to his old engine, which he proceeded to shut down by closing valves and pulling wires off, and then throwing the drive belt off.

Finally everything in the shop wound down to a dead stop, but the siren in my hand was still giving off a low moan. Then he turned around and took a harder look at me and said something like, "Mother of Christ, what the f—k was that?"

I eventually got the thing mounted on the boat and drove it by V-belt from an old Ford Model T starter motor. I used it for years.

To all outward appearances Frank was deadpan serious about everything, without so much as a flicker of a smile, so most people would not attempt to joke with him. In his own way he really enjoyed a good joke but just didn't show it. Otherwise he would never have put up with some of the things I did.

Only a few months after I bought the *Five B.R.*, I encountered my first serious mechanical failure with the new vessel. For tractive power Bob Weld had converted and installed a twenty-horsepower Cletrac tractor engine coupled to an old marine clutch and reverse gear which was too small for the job. Finally one time when I went to go astern the drive shaft broke and left me dead in the water. I found it was quite a jagged break, and providing I went ahead, the propeller thrust would force the broken ends together and I could proceed very slowly without it slipping. So I pointed her in the right direction and, with my fingers crossed, kept her coming ever so gently all the way down the coast. I daren't stop, but kept going till I reached Lund, where I executed a dead stick landing and, with the aid of pike poles and shouting, got her alongside Frank Osborne's float. I expected to buy a new Paragon reverse gear, which would cost about $350, but Frank said, "To Hell with them. They ain't any better than the one you just bust." He'd build me a better one. It would only take two or three days, and he guaranteed it would not cost over a hundred dollars.

At the back of the shop lay the remains of Frank's old red Nash touring car, which was wrecked when he drove off the gravel road on his way back from the Powell River beer parlour. He took the rear end and differential gears out of it and used these as the heart of a reversing gear by putting a drum and brake mechanism on it. The principle was much like that behind the friction bands and planetary gears in the old Model T Fords. It provided me with one hundred percent reverse action. To all this he added a simple clutch which he built up of odd parts. He worked without plans or measurements, but everything turned out perfect and it lasted me the life of the boat. It was now the third day and the job was all but finished Saturday night.

Sunday morning we went to work as usual, but when I got to the shop, Frank was doing something else. He said, "Hell, this is Sunday. It ain't good for a guy to work every day. Today I'm gonna make a cannon for yer boat. Every boat hasta have a cannon." He was quite serious. He already had a short length of bronze propeller shaft in the lathe and was in the process of turning it down into a beautiful replica of an old-time brass cannon. He bored it out with a half-inch drill and then drilled the nipple and attached the grunions. He found a solid teak board from which he fashioned a gun carriage, adding two axles and four brass wheels. He polished it in the lathe using jeweler's rouge, and the whole thing was a joy to behold. It was finished by noon.

He said that this was for real, and all we needed was some gun powder and he would show me how to use it. He sent me over to see Gerald Thulin at the Lund Hotel. His father, Fred Thulin, had owned an old Winchester 44-40, and Frank thought there were still some of the old black powder cartridges around. He was right. Gerald gave me a handful of them, which we then proceeded to unload. Frank was extremely cautious in handling the powder, and felt he had to instruct me in every detail, even though I tried to tell him that as a kid I had made little guns and used lots of black powder when I had the Polish hermit's Swiss Army rifle. But Frank ignored my protests and insisted on lecturing me in painful detail. He even made a cute little brass measuring cup with a teak handle so I could put the precise amount of powder in the barrel each time. He said, "Now, one jigger is plenny, but if you wanna real big bang you could go to one and a half, but no more'n that, y'unnerstan?"

That afternoon we took the little cannon down on the float and fired it time after time until Frank was satisfied that I knew how to handle it. The noise attracted quite a crowd for a Sunday afternoon and Frank suddenly realized he didn't want to be seen by his customers so frivolously engaged, so he left me to it. Also, we had run out of powder. I told him I would go over to Savary and get a one-pound canister that Dad had stored away, and that when I came back that evening I might fire it off a few more times if he didn't mind. He said no, it was okay, but remember—"No more'n one an a half cups, and be damned careful where you point it if yer using any slugs." I said, "Yes, yes Frank, don't you worry. I'll be careful." And then, as an afterthought, he said, "Be sure you tie a rope on so it don't jump overboard on you. It's got a real kick." I promised this would be done. His intentions were of the best, but I was getting a little tired of his treating me like an inexperienced kid.

While I was over at Savary, I had an idea that rapidly developed into a course of action to get Frank's goat. Stored away at the back of the property, Dad had part of a box of twenty-percent dynamite that we had been using for road building. I picked up two sticks, a detonator cap, and a length of fuse, and went back to Lund for an evening's entertainment.

I would make Frank think the cannon had blown up. I would plant the dynamite on the rocks in front of the shop, set it off, and when Frank came down I would hide the barrel of the cannon and show him just the carriage on the rope and see what he would say.

This part of Lund was on a very small bay, with a few houses, the machine shop, Dan Parker's one-man sawmill, Hugo Johnson's little marine ways, and then the Thulin's old Malaspina Hotel and government wharf, arranged in that order, in a rough semicircle. Frank lived in a little house directly above and back of the machine shop, with a steep set of stairs leading up to it. Out in front of the shop, exposed at low tide, was a large pile of rocks that had been left there after the beach was cleared for the marine ways. The shop itself sat on pilings.

Hugo Johnson at work on the Niluht *in Lund, 1921. Engine for the new tug was taken from the wrecked* City of Lund, *right.*

The Five B.R. *tied below Frank Osborne's Lund Machine Shop for repairs to the clutch. Charge was placed just to right.*

View of Five B.R.'s *wheelhouse roof: the siren is on the lower left (with holes.) The starter motor above turns it.*

Two of Frank Osborne's masterpieces: the brass cannon thought to have shot over Lund, and behind it a larger cannon he made later.

Frank Osborne's homemade clutch for the Five B.R..

The tide was out far enough to dry the beach and was still dropping. I placed the dynamite by a rock that was out almost at the geometric centre of the semicircular bay, so the sound of the shot would be well distributed. I covered the dynamite with a bucketful of moulding sand from the foundry, figuring this would be safe and produce no shrapnel. Then I attached a three-minute fuse. Before lighting it, however, I went over to Hugo Johnson, who was taking advantage of low tide to caulk a boat. I told him roughly what I had in mind and asked him if he could move over to the other side of the boat until the charge went off. He seemed to be pleased at the prospect of pulling a joke on Frank, and willingly complied. Then I had one further idea. I knew that Frank, at this time of night, would be listening intently to his favourite radio program, Jack Benny, and he would probably be wearing headphones, so he might not hear the dynamite. I thought I better do something that would ensure getting his attention. His radio was a special one that I had built from scratch to be run entirely off the 32-volt bank of batteries that Frank had in the machine shop, and a pair of wires came all the way down the hill to connect to the batteries. I decided I would slip into the battery room and disconnect the wires the instant the shot sounded. As things turned out, this was overkill.

It was a Sunday evening and nothing was moving. The weather was dead calm, without a breath of wind to rustle a leaf. I double checked. Everything was set. I lit the fuse, ran in, and took up my position in the battery house. With the wire in my hand, I waited. And I waited. It was the longest three minutes I could remember. I was beginning to think I had a misfire on my hands. And then—BLAM! The whole building shook. Loose pieces of glass fell out of the old shop windows, and I involuntarily yanked the wire off as my feet left the ground. The noise it made was appalling, and for the better part of a minute echoes were coming back from the hills around. I fumbled around in the dark and reconnected the wires, then made my way down to the boat as quickly as possible.

Things started to happen. Porch lights came on, dogs were barking all over Lund, and gradually more and more people appeared, carrying bugs and flashlights and generally heading down to the dock area. I looked up to the house and could see Frank out there busily pumping up and lighting a Coleman gas lantern. When he got it going to his satisfaction he reached up and jammed his old black hat on his head and started down the steps with the Coleman sizzling. I didn't have long to wait now, and I'd better have my story

ready. I was on the edge of panic. I think the only thing that saved
me at this time was hearing the methodical plink-plink-plink of
Hugo's caulking mallet resuming its rhythm. At least there was one
person around with his wits collected.

When Frank arrived I was standing on the float holding the piece
of rope with the gun carriage attached, trying to look innocently
bewildered by it all. Frank stuck the lantern in my face and
examined me closely. My eyes were red and watering from nervous
strain, and he took this to indicate a state of partial concussion.

"Are you hurt, kid?"

I assured him I was just a bit shook up.

"What happened to the cannon?" he said.

I showed him the carriage.

"Wurrs the burrel?"

I gave a helpless shrug and indicated I had no idea.

"Which way wuz she pointin'?"

"Out that way where you told me."

He considered that for a moment, then turned, and with his hand
on edge indicated the exact reciprocal direction, across the clam beds
and up to the sidehill.

"She's gotta be thataways. We'll find er."

He started to wade across the mud, searching with the gas lantern.
He went right to the rock and I was afraid he would see evidence of
the blast, but he found nothing.

By this time quite a number of people had turned up. Some were
just inquisitive, some concerned and helpful, one or two quite
hostile. They had not appreciated the cannonading exercise of the
afternoon. Frank got them all organized in the search, and pretty
soon the tide flats and the surrounding hillside looked like it was
crawling with fireflies, with all the flashlights and lanterns. I stayed
on the float to help direct people to the search area, all the while
wondering how far this thing was going to go. Then Brooke Hodson
turned up. Brooke was the Dominion Government telegraph
operator and lineman. He lived about 4½ miles down the road from
Lund, had heard the explosion and come all the way up to Lund in
his pickup truck to find out what was going on. I knew Brooke well,
having worked with him on the line, so I grabbed him by the arm and
took him onto the boat and into my confidence. I showed him the
barrel I was hiding and asked for his help. He said, "Leave it to me;
I'll think of something," and departed with the barrel.

Among the many that showed up at the scene was Alec Johnston, the Provincial Police constable stationed on the police launch *PGD2*, at that moment moored about a mile across the harbour in Finn Bay. The explosion shook him out of his bunk and he rowed across the bay to find out what was happening. I thought I'd better take him into my confidence too, and he was good enough to keep the peace. I don't know how much longer the search went on under Frank's urging, but the tide was still dropping and the *Five B.R.* was in danger of going aground, so I took advantage of this to make my exit and tie up over at Finn Bay, but only after promising Frank I would be back in the morning to resume operations. He said he intended to get all the school kids out after class the next day. I could see this leading us further and further into trouble, but next morning it was all nicely taken care of, thanks to Brooke Hodson.

Brooke still had the barrel in his pocket Monday morning when he met Dr. Lyons, who had just come up from Powell River to carry out a routine medical inspection of the Lund School. Brooke saw his opportunity, filled Dr. Lyons in on the details of the story, and asked him to plant the barrel in the schoolyard on his way in. When the kids came out for morning recess it was no time at all before one of them discovered the barrel. Everybody had heard about the episode by then, so the boy took it immediately to the teacher, who appreciated the seriousness of the situation and sent two boys straight down to Frank Osborne with the barrel. Frank simply couldn't believe it and went back to the school with the boys so they could show him exactly where they found it planted muzzle down in a patch of dry sand.

I met him when he came down to the boat, holding the cannon barrel gently in both hands and staring at it, shaking his head in disbelief.

"That school hasta be half a mile from here, and them trees must be two hunnert feet high! For Chrissake, how much powder didja put inut? Didja put more'n I toldja? Didja use the cup I made you?"

I admitted that I may have got carried away and put in more than he said, but he still couldn't accept that this would have wrought all the havoc it did.

"Wadja put inut fer a slug?" he demanded. I had made up a long and, I hoped, plausible story. I told him I couldn't readily find a suitable slug that would fit the barrel the way I thought it should, so I had decided to go first class and mould one out of babbitt-metal. He had been using the forge the day before and had left some babbitt in the ladle after pouring some bearings. He wanted to know what I had used for a mould. I admitted that this had been a bit difficult until I got the bright idea of using the gun barrel itself as a mould, first putting a wad of wax paper and moulding clay on top of the powder charge.

That did it. He threw his hat on the deck and stamped on it and shouted, "For Chrissake, you can't solder them things up and expect em to stand it! It's a wonder you didn't kill yerself, and if she hadn'ta bin built strong you wouldadone!"

"Lookit this," he said, and took a pair of calipers out of his breast pocket and carefully gauged the barrel at several places in the bore and then said, with some pride in his voice, "She might be sprung a little at the breech, but the rest of er's as good as the day I made er!"

Then he allowed himself to reminisce a bit, shaking his head in wonderment the whole while.

"My Gawd, what an explosion! It shook the ground. Two of the old lady's dishes fell offa the rail and bust, and you couldn't see across the room for dust. And you know? It stopped the goddamn radio right when I was listening to Jack Benny. Was a coupla minutes before I could get 'er rolling again. Hadta take my jackknife and jist when I touched er to one a them condenser plates, she started up again. That's a little trick you oughta remember, it might come in handy when yer out fixin'. Here's your goddamn gun, I got work to do!"

By Monday afternoon things began to get awkward. Some knew the story and some didn't. Some were anything but pleased to have their sabbath disturbed. The schoolteacher, who was a serious type, delivered the class a long lecture on the dangers of explosives generally, and cannons in particular, and apparently used my name quite freely throughout. I left that afternoon for the north and didn't return to Lund for at least a month. But to my amazement, the story of the cannon preceded me wherever I went. It was spread by the purser and crew of the Union Steamship boat, and all the towboat crews that called at Lund heard and spread the story. There were many versions. Some had it that I had loaded the cannon with two sticks of dynamite and blew the whole machine shop apart. As for Frank, he delighted in telling the story till the day of his death. I saw him many years later and the first thing the old man said was, "You remember that goddamned cannon I made? Have you still got it?"

He never learned the truth, or if he did, he preferred to remember it his way. Bless him! Bless him!

THIRTEEN
A Stormy Honeymoon

LIVING FULL TIME aboard a floating workshop was an excellent proposition from the standpoint of my business, but it was lonely. All my close friends started getting married. Hob Marlatt, who had become my closest friend during this part of my life, married a very nice girl and went to work for a shipping company in Vancouver. I used to go down to visit them in their little apartment on Pender Street. Allan Mace got married too. Glenys Glynes and I had been seeing each other again, and marriage began to seem an obvious move. It wasn't terribly well thought out, but she was a very sensible type and I needed companionship on the boat. My family didn't object (she was English) so it was more or less expected, but when we did decide it was practically overnight, the week before Christmas in 1937.

I was facing a particularly busy schedule of promised calls, and actually had some deliveries of new radios to make. Business had been particularly good. I was on my way to Vancouver with Leo Kent, a machinist from Cortes Island, along for the trip. He had to go to the Columbia Coast Mission hospital at Pender Harbour to have all his teeth pulled, and when he got that far he asked to come all the way to Vancouver, where he planned on looking for some machine parts in the auto wreckers. In those days I was always glad of company.

The dentistry episode in Pender Harbour is worthy of mention. There was no dentist, but the doctor, Keith Johnson, fulfilled the function. They did have equipment — of sorts. The dentist's drill was an arrangement of pulleys and belts driven off the rim of a bicycle wheel. It took two people to operate. The hospital orderly would have to climb up on the seat and pedal the contrivance while the doctor handled the drill. Anyway there was no cause to use it on Leo. His teeth were well past that stage. After one look, Dr. Johnson said they had to come out — every one of them. Leo told him to go ahead.

It must have been quite a performance. Leo came down to the boat wiping blood off his mouth and trying to learn to talk without teeth. He made himself understood, though. He sputtered, "Fat's de last goddamn time I'll let that sunnova-bitch pull a tooth outa my head!"

I cleaned up the last of the pre-Christmas service calls in Pender Harbour and then we headed for Vancouver and tied up in our usual spot alongside the Pacific Coyle Navigation floats at the foot of Cardero Street. Then there was lots to do. Up to Radio Sales Service to replenish my stock of tubes, batteries, and spare parts, and to pick up the three new radios I had promised for pre-Christmas delivery. In addition to this I had promised Dad I would do some special shopping for him — Pumphries coffee beans, malt extract, and similar items not normally available by mail order.

And then, of course, the court house. Glenys had reminded me that they required three days' notice, in the clear, for the marriage licence. It could just be squeezed in. Then I had to get hold of Hob Marlatt to be best man. Glenys's sister Alison was to be bridesmaid. This was already a bigger crowd than we wanted, but the ceremony did require two witnesses. The great day came, and it was also the deadline for departure for upcoast points. It was touch and go. I was up town picking up the last few items to go aboard with only minutes to spare. I was to meet the others on the courthouse steps.

I had only one good pair of pants with me and in anticipation of the event I was wearing them. Unfortunately during the morning's activity I had spilled some oil or something, and they were in a mess. They were light grey flannels, and it wouldn't wipe off. I had to get them cleaned. There was one of these small places that advertised *Dry Cleaning While You Wait.* I headed straight for it, and sure enough the Greek gentleman was pleased to accommodate. He said, "I can't do it unless you take them off, but just sit on that chair, no one will notice, and I'll only be a few minutes."

Now I began to understand the full meaning of the sign on the door— ... *While You Wait.* You wait, and wait... Every time he showed up I implored him to hurry. I even told him I was going to get married. He responded by saying that was fine—a good idea—he liked to see young people get married. The only proper way to do it. He didn't go for this business of just jumping into bed—too much of that sort of thing going on nowadays. He was old-fashioned. I said, "Look, I'm sorry, but I really am getting married, and if I don't go right now I'll be late!" He said listen, if it's that important you wouldn't want a poor job on your pants, and anyway they won't be dry yet, they're in the oven right now. Keep your shirt on.

When I finally got into them they made me jump, they were so hot. I paid him his fifty cents and I was still doing up the last buttons when I ran out of his shop and never stopped running till I got to the Court House, which was about five blocks. I actually got there just as we were asked into the registrar's office, but the others were getting pretty worried. All Glenys said was, "What happened to *you*?" We hadn't been married long enough for her to be more pointed, but in later years I often heard her remark, "He was even late for his own wedding!" During the ceremony, which fortunately was brief, my eyes were running, and to begin with Glenys thought it was due to sentiment and was very touched, until I explained that after I stopped running the fumes from the overheated cleaner's fluid were coming straight up into my eyes. When Hob got a whiff of it he thought I had been into a bottle already. Anyway we made our way down to the *Five B.R.*, said our goodbyes, and before long were on our way through First Narrows headed north.

I don't know when it had first occurred to me, but it became obvious that, as well equipped as the *Five B.R.* was, its layout and design was not entirely suited to the requirements of married life. There were five bunks aboard, but they were all single. Leo Kent, who was returning with us, was an understanding sort, and came up with a very clever design to convert the one bunk in the saloon into a double. It consisted of replacing the bunk board with a series of alternating slats, half attached to the wall and the others to the front board, so the whole thing could be slid out to double width, supported by a couple of chains from the deck head. The back cushion slid down and formed part of the mattress. Presto! Clever, but it required some intricate carpentry and a lot of lumber, all of which was loaded aboard when we left town. Leo undertook to do the necessary construction once we got under way, while I steered and Glenys found her way around the galley. We met a northwesterly that freshened as it got dark, so we decided to go in to Snug Cove, Bowen Island, for the night. Instead of mooring alongside the other boats, I decided to anchor well off so Leo's hammering on the bunk construction wouldn't disturb the other people.

We were up most of the night, but he did finish the job. We weighed anchor at first light and travelled all day till we arrived at Cortes Island and disembarked Leo, minus teeth. We called at Refuge Cove, where we arranged to return on Christmas Eve to spend Christmas day with Jack and Rose Tindall and eat their turkey. They also were just recently married. It would have been quite impossible for us to plan to spend the night with Dad at Savary Island because of the lack of shelter and prevailing weather at that season. I note from the log book that for about three weeks we had frequent strong gales, snowstorms, and it was bitterly cold.

December 24 we arrived at Heriot Bay, where I installed a new Stromberg-Carlson radio for Francis Dickie the writer. By this time he had despaired of getting the new set before Christmas. After making Dickie happy we had three more calls, including a new set for S.K. Marshall the Greek scholar, who had a large English-style house in Evans Bay. Our last pre-Christmas commitment fulfilled, at 17:30 we left Evans Bay bound for Refuge Cove and supper. It was snowing, with a biting northwest wind, and pitch dark, of course.

One hour later we were heading into what the locals refer to as a Bute wind. It is an Arctic-air outflow that comes straight down Bute Inlet, and this one was a dilly. We found ourselves heading into dirty four-foot swells, with the spray starting to freeze on the handrails and rigging. Visibility was almost nil, and we were making very little headway. We shouldn't have been out there at all, but I wanted to keep our appointment for Christmas dinner with Jack and Rose.

We were just abeam Bullock Bluff, the northern tip of Cortes Island, and nearly to the point where we could put the wheel over and head down Lewis Channel with the wind astern, when, without warning, the engine stopped dead. We were about half a mile from shore. The vessel immediately fell off wind and lay broadside, rolling madly. I raised the hatch and jumped down into the engine room to try and find the trouble. Everything checked okay so I grabbed the starting crank and spun it over — nothing. I checked all the obvious things, gas, spark, distributor wires, but everything appeared in order. I spun the crank again — again nothing.

I climbed out and sized up the situation. It wasn't good. We were making leeway while we rolled, and already had used up half the distance to the rocks of Bullock Bluff. The wind was screaming, and we were rolling viciously. It was hard to hold on. Given another hundred yards we would have made it clear of the bluff, but the tide was ebbing, and the current was carrying us south, so there was no question we would pile up on the bluff in a matter of minutes. Our situation was most uncomfortable — desperate, in fact. So I did the only thing left to do. I scrambled out the wheelhouse door and onto the icy foredeck and lowered the anchor. The term *lowered* hardly describes it. I just kicked off the dog and released the brake and let the windlass scream out to the bitter end. It carried 175 feet of wire cable and chain, and an 85-pound navy anchor. I could only hope that the anchor might take hold before we hit the beach, but the water is deep off the point and it would be touch and go. However, I had done all I could, so I slithered back into the wheelhouse and down into the engine room, and started desperately to crank again. All this time poor Glenys was just hanging on like grim death and saying nothing. What an introduction to boating — and married life!

I think it was at this point that I decided to pray. I seemed to recall that it was what other people had been known to do under similar circumstances. But I was not a very good hand at that sort of thing. I wracked my mind but in spite of being brought up by a religious-minded father and named by a clergyman grandfather, I couldn't think of any words. What would the Governor think of me now? Then I wondered just how important the words actually were, so long as the intentions were there. If the scheme could work, I was more than willing to give it a try, words or no words. Then words came to me — silent words, but they surfaced in a rush: "Break, break, break, on thy cold grey stones, O Sea! And I would that my tongue could utter the thoughts that arise in me." Not out of the prayer book but I was in no position to be choosy.

I felt the vessel suddenly swing around into the wind, and the rolling eased. The anchor had caught, and only just in time. I rushed out to the stern with a flashlight to have a look, and I'll never forget the scene that greeted me. Our stern was within ten feet of the steep rock cliff. I could have reached it with a boathook. It had large icicles hanging down where the spray had frozen. Our stern was heaving up and down five or six feet and as the breakers hit the rocks the spray came right back into our cockpit. I rushed back to the engine room and continued my frantic search for the problem, but still no clues. After a while I began to worry about the situation at the stern, so I ran back and had another look. We seemed to be closer — too damned close — and then I realized that the tide was falling and the anchor cable would be slackening and we would be getting continually closer. And then suppose the anchor was to slip just a notch? That's all it would take.

We would be thrown broadside against a solid rock cliff, and in those waves the vessel would be reduced to matchwood in a few minutes. Climbing ashore on those rocks in those waves would be virtually impossible. And even if we managed to get ashore, then what? It would be over seven miles of steep rock and forest to the first habitation, Emery and Clayton's logging camp at Junction Point, and being all wet I doubted we would survive many hours in that biting cold. The chances of attracting the attention of any passing boat were about nil. This was Christmas Eve and everybody but us was home.

Our only hope would be the eight-foot pram dinghy which we carried atop the main cabin, tucked against the funnel upside down. We had a small boom and winch on the aftermast for normal handling of the dinghy, but in those thrashing seas it would be impossible to manipulate, even if we had time. We would have to heave the boat over the rail by hand and try to jump in before it was carried away — being careful not to lose the oars. Then to survive the steep waves and backwash in the vicinity of the wreck and row against the wind till we were clear of the bluffs, to make our way down Lewis Channel in the dark, snow and wind for seven miles?

I rushed up to the foredeck, and very, very gently took up a few feet of cable on the windlass — always risking the possibility that in doing so I might dislodge the anchor. It was touch and go, but it worked. We now had a few more feet of clearance, so I returned to the engine room once again. This time I must have touched the right thing, because I gave the crank one pull and the engine started. To this day I don't know what the trouble was, and it never happened again. But we were not quite home free yet.

Savary friends. Standing: Hob Marlatt, Alan Mace, me. Seated: Alison Glynes, Marie MacDonald, Glenys.

The honeymoon cruise begins. From left: me, Glenys, Alison, Hob.

The Five B.R. *at Refuge Cove.*

Jack Tindall and Margaret.

I jumped out of the well and gently eased the clutch ahead, gradually increasing power till the ship started to move slowly forward against the gale-force wind and nasty seas. I just kept it coming slowly forward till it lifted the anchor off the bottom. I was taking no chances, but kept creeping ahead, knowing that if I tried to increase speed there was the danger of the anchor cable being drawn into the propeller, particularly with the vessel pitching like it was. I think it may have been half an hour before I was dead sure that I was clear of the bluff with the whole of Lewis Channel clear and astern of us. Then, and only then, I turned the wheel over to Glenys to keep the ship's head into the wind while I got out on deck and winched in the anchor. We were free to make our turn downwind and into a snug moorage in Refuge Cove.

FOURTEEN

Living Aboard

WHEN WAR BROKE OUT Hob joined up immediately. He took his flight training and I saw him once more when he came up to Savary. He had his wings. He was all fired up about planes, describing the ones he'd flown and adventures he'd had in training. I was taking colour movies then, and I took movies of him showing off his uniform.

Two months later he was dead. He flew a Mustang into a cloud full of rocks in Scotland.

I travelled to Vancouver the first chance I could and went to the recruiting office on Seymour Street for an interview. A very serious young chap behind the desk wished to know my business. I told him I would like to join up, and what was the setup? He said, "What service?" I said, well, I'd like to train as a pilot, or failing that I'd like to get into air force communications because I'm a radio technician and I understand there's a great need for that. "Education?" he said. I told him about being qualified to attend any high school in BC. "Really?" he said. "Unfortunately, we're not taking anyone with less than three years of university." The only thing in the air force that didn't require a higher education was tailgunner, and even for that privilege you had to go on a waiting list. Then he loosened up a little and offered me some fatherly advice. "You say you're married, that you've got a little business—I suggest you go back to it. This isn't your kind of war." He gave me quite a lecture. He was of the opinion that Hitler was going to be taken care of entirely by individuals having at least three years university, and in very short order at that. Not your kind of war, he said. Righto, I thought. I went back up the coast and didn't try to enlist again. Later, of course, they were taking any warm body they could lay their hands on.

I kept on going up and down the coast. Living on the boat went fine. The first year we kept accurate track of all our expenses and it came out to $24 and some cents a month. This included all our groceries and heating (but not boat fuel) and some entertaining. We didn't drink. I kept a bottle of Johnny Walker Red Label on the boat and it lasted two years. But Glenys was a good cook and we often had people down on board the boat. Glenys had never lived on a boat before but she took to it well, and we made a lot of friends. A lot of places where we'd go, just having a woman on board was very popular. The ladies would come down and say, oh, your wife must come in. We're doing this that and the other, knitting and all the rest of it.

We were comfortable, and the coast kept us from getting bored. We would begin around Savary, go over and make calls at Whaletown and Gorge Harbour on Cortes, drop down to Manson's Landing and Blind Creek (which is now Cortes Bay), then Squirrel Cove and Seaford on the other side, and even up at the head of Von Donop there were people living then.

After the logging pulls out, somebody decides he won't go back with the camp, he likes it there, he makes a garden and goes fishing and stays. That's the way the whole coast was settled in those days. I could name twenty of these characters in Desolation Sound alone. Old bachelors working away by themselves on little handlog shows. They were so ornery nobody would work with them, and they'd been there forever.

It was a hot-spot, but as you went north you had the whole thing repeated around Jackson Bay, Topaz Harbour, Forward Harbour and Port Neville. Port Neville was a real centre for settlement. The Hansens ran a store and post office and there were a lot of settlers further in. The flats at the head were all fenced off in farms, and there was lots of cattle wandering around. Then around Havannah Channel, Port Harvey, Chatham Channel, Minstrel Island, and Clio Channel there were an enormous number of people, mostly loggers. At the heads of the inlets they would go up the rivers sometimes twenty or thirty miles. Toba had a whole settlement you never hear about, twenty miles up the river. That one was pretty well finished by my time, but I knew some of the people who'd grown up there. At the head of Knight Inlet you had the Stanton family and up Kingcome Inlet the Hallidays, who ran a considerable cattle ranch until very recently. There were a lot of them, and some of them dated back before the turn of the century.

Over on Vancouver Island, of course, the thing was multiplied enormously, but I didn't know that area. It was accessible by road from Victoria pretty early on, and by rail as far as Courtenay, so it was far ahead of the mainland side. I didn't do much business on Vancouver Island south of Sayward. There was a settlement back up the Sayward Valley and I went in there. Then up to Alert Bay, which was an important stop. Port MacNeill was not much in those days, only one big logging camp, but Englewood I used to go into.

Going on up, there'd be the Laviolettes, who ran the hotel and beer parlour at Echo Bay, and the Dunseiths, who ran the floating store next door at Simoom Sound (a town entirely on floats which moved at least once), then O'Brien Bay and Sullivan Bay. Across from that was Claydon Bay, the big Brown and Kirkland camp where I did a lot of business. Then, around the outside and up Wells Passage into Schooner Passage and through the Nakwakto Rapids into Seymour Inlet, a whole new world opened up. In this one inlet there are 1,100 miles of uncharted coastline—including Nugent Sound, Belize Inlet, Seymour Inlet, Mereworth Sound, Allison Sound and a mess of lagoons and salt lakes, up and back, up and back, and at the head of each a big river. All the water has to flow out through Nakwakto Rapids, which is only 75 yards across, so the ebb currents are much stronger than the flood. There were hundreds of people tucked away in there. These were not settlers; these were strictly loggers when I first started going up in 1937. Oscar Johnson was our favourite. He'd say, "My name is Oscar Yon-son and I live yoost a ways up Nu-yent Sound." Dumaresq's were the largest camp with, I guess, 150–200 men. There were eight or nine camps of good size.

Further than that I didn't go. Seymour Inlet was my furthest north. But it kept me busy, between there and Pender Harbour. I felt I had to visit Pender Harbour about once a month because there were so many people in around there, and in those days they had no road connection to Vancouver. I printed a list of all my regular stops running down the left side of my "A.J. Spilsbury, Radio Technician" letterhead. This is it:

Radio service launch "Five B.R." calling at—

Blind Channel
Bliss Landing
Blubber Bay
Bold Point
Bute Inlet
Campbell River
Chonat Bay
Church House
Cortez Island
Cracroft
Egmont
Elk Bay
Frederick Arm
Granite Bay
Heriot Bay
Jackson Bay
Jervis Inlet
Lang Bay
Loughborough
Lund
Menzies Bay
Minstrel Island
Nodales
Palmer Bay
Pender Harbour
Phillips Arm
Port Harvey
Port Neville
Powell River
Quathiaski Cove
Ramsay Arm
Read Island
Redonda Bay
Refuge Cove
Rock Bay
Roy
Savary Island
Sayward
Seaford

Seymour Narrows
Shoal Bay
Squirrel Cove
Stag Bay
Stillwater
Stuart Island
Surge Narrows
Theodosia Arm
Thurlow
Thurston Bay
Toba Inlet
Van Anda
Vaucroft
Waiatt Bay
Westview
Whaletown
Etc.

I guess if that's a list of fifty names, there's not twenty of them you could still send a letter to. One of my favourite addresses was Roy. It was supposed to be the shortest address in the British Empire. We had a customer there named Harry Blue, so his mail used to go out H. Blue, Roy. And that last one, Etc., was pretty important. We probably did more business with it than any other one place.

We were by no means the only boat carrying on business on the coast. In those days there were quite a number of "store boats" and agents of one kind or another with much the same itinerary as ours. Mac & Mac, the big hardware concern, operated the *Sundown*, a large converted seine boat that carried a wide selection of hardware. Daniel Wetmore and his wife operated another hardware boat, the *Marshall Wells*, for the company of that name. George Bradshaw and his wife sold everything from Watkins gloves to Barrett roofing shingles on the 55-foot *Adventus*. The *Arrawak I* out of Pender Harbour carried men's clothing and dry goods. And many others: seagoing dentists, hairdressers, and insurance salesmen came and went.

We got to know many of them quite well and those who weren't directly competitive would pass information back and forth concerning prospective customers. One fellow who was particularly helpful in passing along good leads to me was Lorne Maynard, a World War I vet with one leg, who lived at Blind Bay. Lorne had a

Provincial policeman Roy Allen at Big Bay with a couple of the area's characters, Jud Ingles and Tommy Thomas.

The Five B.R. *at Castle Falls in Teakearne Arm.*

The Wilcock store at Stuart Island.

Booming Ground at Knight Inlet, *pastel.*   SPILSBURY

nice little boat named the *Therma,* which was originally equipped with a small steam engine and later converted to a small Universal 12–15 horsepower gasoline engine. He had a regular contract to take the boiler inspector up and down the coast year after year to inspect and test all steam boilers. In those days there were hundreds of steam donkeys used in the woods so it would take him the better part of a year to cover them all.

Once Lorne told me that the next time I was in the Minstrel Island area I *must* call at Soderman's Camp in Burial Cove. They needed radio work and he had promised to send me in. I had often heard of the Sodermans and I had taken pains to avoid their floatcamp. With Glenys along I had an additional reason: it was said they would not permit a woman aboard, even as a visitor. Oscar was a Swede and okay, but his wife Sidney had acquired a fearsome reputation. How they came to be together was part of coastal legend.

In earlier times Oscar Soderman, an old-time logger with a rather ramshackle camp, was noted for his periodic drinking sprees—work like hell for six months or so and then sell the boom and go in to town and spend the works on booze. Then sober up and start all over again from the bottom up. At a certain point he went contract logging for Pacific Mills up near Ocean Falls and, aside from his drinking, was doing quite well. In those days Ocean Falls was a booming coast town with a population of several thousand. It had a red-light district that prospered along with it, but when the authorities found out they ran the girls out of town.

But not very far. Right across the inlet, about two miles from town and just outside city limits, the madam, who had managed to set aside a tidy fortune, established a new place of business. The "house" was supported on pilings. The location thenceforth became known to history as Pecker Point. Before long the enterprise began to pay off handsomely, for Madam proved to have excellent business instincts. In fact the enterprise became so prosperous that she decided to bring things up to date and establish regular working hours. This included closing the joint on Sundays to give the girls a day off.

Shortly after this innovation the big steam tug *Sea Lion* came into port to pick up a barge load of pulp, and while waiting for the barge to load, decided to avail themselves of the services across the inlet. They steamed over to the bay and moored alongside the "house." But it was Sunday, and the new regulations were in effect. Captain Andy Johnson pleaded with the madam—but to no avail. The doors were closed and that was that. But Johnson was a hard man to stop once he got steam up. He got his boys to launch a lifeboat and pass a towline right around the whole clump of piling, and then the *Sea Lion* steamed away from the dock and took a strain on the tow line. The whole house moved slightly to seaward. No response. The doors remained locked. He rang down for a few more revs. The house began to creak and groan and the floors tilted. Johnson got out his megaphone and advised the madam she had one more chance or he'd ring down for full steam ahead and dump the whole shebang in the chuck. The madam opened the doors and waved them in. That was the end of Sunday closing.

Along about then Oscar Soderman sold his boom and came to town for a real drunk. He just didn't know *how* drunk he was going to be. After he met the madam he remembered nothing until he woke up in a Vancouver hotel and she told him they were now man and wife. He remembered nothing, but there was no mistaking his signature on the certificate she was waving around.

Financially it was a great success. She sold the house on pilings and invested in his logging company, applying her keen management sense to good effect and putting an end to Oscar's big binges. After several years of joint management, the Sodermans' camp proved itself financially successful, paid their bills in cash and had no debts, but a tough reputation. They were reported to have *three* crews — one coming, one going, and one working. It was a merry-go-round. Sidney ran the camp and nobody managed to live up to her expectations for long. When the hiring agency sent up a batch of new crewmen on the steamer, she would meet them coming down the gangplank and give them a quick once-over. If a hopeful chokerman looked a bit on the youthful side she'd holler, "Back on the boat, you! We want men in this camp, not boys!" If another looked in need of a haircut she'd roar, "Hey, Goldilocks, you can't work here like that, you'll be pulling your hair out of your eyes all day! Git!" Everyone was scared of Sidney. Oscar had absolutely no say.

It was Sidney who wouldn't permit another woman in camp. If I hadn't owed Lorne so many favours I would never have called, and as it was I was fearing the worst. I told Glenys some of the background — not all. Glenys was very straight-laced and wouldn't have let me go near Burial Cove if she knew the whole story. I suggested she stay out of sight until I was sure of the situation.

It was mid afternoon when I brought *Five B.R.* carefully alongside the main float. All of the crewmen including Oscar were in

"Etc. BC" — *one of the many nameless floatcamps served by the* Five B.R.

Admiring the view up Bute Inlet.

Mount Churchill, Jervis Inlet, *pastel*. SPILSBURY

the woods. I could hear the steam donkey tooting and the rumble and squeaking of the rigging. Sidney was standing on the float to meet us. She instructed me to bring the boat in through the gap and moor it behind the main float where we would be sheltered if the wind got up.

"You might as well be comfortable, there's several days work for you. Tell your wife to freshen up in my house. We'll be eating in the cookhouse at six."

She was obviously expecting us and could not have been nicer. She was of medium height, about fifty, with wisps of greying brown hair straggling out from under a red cotton bandana pinned around her head like a turban. She wore no makeup. A flowered silk blouse showed she had been built right the first time. From there on down it was men's trousers tucked into knee-high laced boots. She wore expensive jewelry. In some respects she reminded me of my mother. She was very well spoken and never once used an off-colour word in our presence. It was the reverse of what we expected—but then she was on her good behaviour for us. According to all reports this did not hold when she was talking to her loggers.

The floating camp was quite usual—a cookhouse, several bunkhouses, machine shop, store room. The difference was a large, completely furnished private house with about seven rooms and all conveniences. It was not occupied. Sidney slept in a small cabin alongside and Oscar slept separately in another cabin at the opposite end of the camp. They met only over meals in the cookhouse.

Once the ice was broken Glenys and Sidney got along like a couple of long-lost buddies. Sidney insisted Glenys use the house and all its facilities—refrigerator, washing machine, bath. Glenys responded by cooking up a large batch of caramel fudge which was very popular. We stayed two days. Sidney kept thinking up more things for me to do. She already had the most expensive De Forest Crosley Radio which I got working in top notch condition, and then she had me wire the whole house for speaker jacks so each separate room had its own speaker and volume control. All wiring was concealed. This meant that I had to move furniture and run the wires along the baseboards. When I came to the master bedroom I moved the bed out from the wall to do this, and underneath I found over three hundred dollars in coins and bills, just laying there. I swept it into a pile and then went out to tell her about what I had found, and she said, "That's okay kid, just sweep it back under when you're through. There's no way I can spend that kind of money up here."

There were heavy drapes on all the windows, and pinned to the drapes were literally dozens of cheques, not cashed. They were made out to Soderman's Log Company, in amounts up to and exceeding five hundred dollars in payment for boomsticks and other materials that various people had bought. She said she had no use for them, and just left the cheques on the curtains. As nice as she was, we couldn't help thinking that we were sitting on some kind of a volcano that might erupt at any minute, so it was with considerable relief that we finally bade her farewell. They were not around too much longer. I think Oscar died. Years later I saw a note in the Vancouver *Sun* reporting Sidney's death, and noting her bequest to the University of BC for a large sum of money.

I still used Savary Island as my home address. Customers would write me there or send one of the reply cards I'd had printed up, Mother would look at my calendar to determine where I'd be next and forward my mail there. It would go down to Vancouver on the Friday boat, then it would take the Monday boat for Rivers Inlet and in ten days or so it would find me at the Bellham store in Allison Harbour, informing me I was wanted urgently at Halfmoon Bay, three hundred miles away. Two months later I would show up, and they'd say, "We gave up waiting." It was very clumsy, and quite frustrating for someone in the business of instantaneous communication, but of course they had closed down all the ham stations at the outbreak of war. I did have the police network, but I couldn't use it for business. Later on there was some relaxation because the telephone company put in its ship-to-shore service. The main frequency was 2142 kHz, and the tugs and the fishermen were on 2366 and 2558. I was then able to get a commercial license for the boat and that's when I got my first marine transmitter. Hepburn, my ham friend in Victoria, built it for me.

Ronnie, our first child, was born in 1940. We had a choice of going to Rock Bay for a hospital or to Pender Harbour. Pender Harbour was by far the best and so when time came due we worked our way down into that area. The doctor there, Keith Johnson, became one of our closest friends. He was a hell of a swell guy. It seemed that he would put down roots in Pender Harbour, and he did for a while, but he had problems we didn't realize. He had wanted desperately to join up but wasn't accepted for medical reasons. He did one tour overseas as a volunteer but then his brother was killed, and he took that very badly. The poor devil ended up killing himself.

One benefit of using the mail was that Dad kept all my letters in a file, which provides me with a comprehensive record of everything that went on in my life for the first five years after I left home. The first one is dated December 10, 1940. We had just arrived in Pender for Ronnie's expected nativity:

Dear Dad,
Just a line to corroborate the phone call, as it was not a very good connection. We left Lund at 00.45 Sunday just as soon as the wind switched to the west, as I quite expected it to return to the SE harder than ever by daylight. We reached Pender by 6.30 with a fresh westerly and a fairly heavy sea all the way. It has been calm and sunny ever since, and to top it all off Doc Johnson says he doesn't expect any developments for two weeks yet by the look of things.
However we are sticking around just in case! No other news at all. Have a great deal to do — Boat just blew, must beat it for the other end of the Harbour so cheerio and love

Jim

There was always lots to do in Pender, and I had to turn down midnight calls from passing tugs in order to stay at my post. But I did have time to write letters every few days to keep my parents posted on the vigil:

Dec. 16
Thanks for the letter, and all the mail received OK on Sunday. There is nothing to tell you from this end. Glenys is well. No developments. We have been quite busy since coming down, and haven't had much chance to catch up on back work so far. The message you heard was probably the *Le Roi* trying to date us up for 5 A.M. one foggy morning, but I didn't bite. Another tug, the *Standby* came in looking for us yesterday. Just had a new outfit installed and I had to work on it for a day to tune it up.
Mr. Greene [the Reverend Alan] just dropped down to the boat to say hello, and sends along his best wishes to you both. He has been away doing a lot of lecturing and is

still on the run. John Antle is also hanging around Pender Harbour with some boat he sailed out across the Atlantic. It's a typical, useless kind of a European product [the *Reverie*] with an engine that won't work. He also has an English radio on board that won't work, and wants me to look at it, so I suppose I will have to crawl aboard the blessed thing sometime, although I told him there is no chance of my being able to do anything for him and his radio as it doesn't employ any standard tubes etc. and probably won't work over 25 miles any time. Wally Smith [engineer of the hospital ship *Columbia*] says he wouldn't take a chance in crossing the Harbour in that darned boat let alone venture outside. Antle doesn't deserve to be alive to tell the story.

Must close now as it is getting late, and Glenys is hurrying me so she can get the table cleared for supper. I wish the weather would get worse or something. There hasn't been a stir on the water since we came and I feel it is a waste of good calm weather to be in here all the time. Affectionately—

Dec. 28

Many thanks for the flowers Mother. They arrived in good condition, nicely packed. We still have the fern and a few of the carnations left. Hate to see the last of them go. Nothing at all to report yet. Johnson says everything is in good order and ready to come as soon as it decides to, which might be any day. In the mean time we spent Christmas on the boat.

Our chief excitement here was when the *Lady Cecilia* ran on the rocks southbound last Tuesday. We were tied at the float at Irvines at the time and had just got on the boat when the steamer backed away from the dock. Heard a noise like a truck backing into a lot of empty gas drums and looked out to see the *Cecilia* with her stern up among the shacks on Indian Island. Apparently the engineer continued in reverse when the signals were ahead. The boat was loaded with passengers headed for town, and was well ahead of schedule. All expecting to reach town in time for supper Xmas eve! The captain sent out a call for help on the radio. Cape Lazo answered and

Lorne Maynard's Therma *tied up at the Soderman floatcamp, Burial Cove.*

Brown and Kirkland's railway camp in Ramsay Arm.

Olas Lee and his son at their homestead in Agamemnon Bay.

Indians gathered for gambling game of lahal (players seated in facing rows at back) in Quathiaski Cove.

sent out a call for any tugs but of course they had all gone into town for the holidays and the *Five B.R.* was the only boat to answer the call. We went over and at their request took a line and tried to pull them off. Another small fish-boat assisted, but of course it was quite futile. We couldn't even pull the sag out of the heavy hauser they gave us. They got the passengers ashore by gasboat and also the mail and baggage. They lowered their lifeboats after much difficulty, got an anchor out in the bay with the help of a gasboat, let all steam down and spent the next 22 hours where she was. That night at low tide she looked pretty dismal with no sign of life, and of course no lights. Her foredeck was awash. In the meantime the *Venture* ambled in to see what all the noise was about, and after wasting about an hour trying to pull her off, went over to the dock and took the hungry passengers to town. The following night the *Salvage Chieftain* and two other tugs were there to tow her away. She refloated herself on a 13.4 ft. tide and they towed her out in wild semicircles as her rudder was all to one side. I believe they eventually had to tow her to town stern first. Love —

Dec. 30

Have just come down from the P.O. after struggling with the telephone for an hour and finally got through to you. I don't know now whether you could hear though as it was a very poor connection then. I also sent a cable to Aunt Bella. Thanks for the suggestion Dad although I intended to do so in any case and was amazed how cheap it was to phone England.

Concerning RONALD WILMOT. He is eight pounds four ounces of boy. (Most of that is lungs, I think.) He has dark curly hair, and is audible for a radius of about 200 yards from the hospital with all doors closed. If he goes on like that I'm afraid they will kick him out for keeping the rest of the town awake. He arrived at the convenient time of 3 A.M. this morning. Like his Pop I guess he likes to travel at night.

Poor Glenys had a pretty bad time while it lasted. It was necessary to operate finally in order to shorten the time as she got very weak towards the end and the doctor said it would have been too hard on her to let it go on. I stayed with her until about 10.30 last night till I couldn't stick it any longer, and at 3.30 this morning the doctor came down to the boat to tell me the good news and we went right up and saw them both. Glenys was out of the ether by then and I was able to talk to her. Johnson seems well pleased with everything and there is every reason to expect her to do well. He is very good, and cheerful about everything—assured me yesterday that there was no need to worry as he had "never lost a father yet."

Your telegram just arrived. Thanks so much. "Old Bill," the wharfinger from Irvines Landing, brought it down to me and then added his own congratulations; with such genuine earnestness incidentally that I felt obliged to dose him with about a mug full of 30 O.P. rum and send him groping back to his post. I hope he got there in case there are any more wires.

Jan. 17

After getting a favourable weather report this morning we left Pender Harbour as per schedule. By the time we reached Welcome Pass the southeaster started to freshen again and we had it a little sloppy all the way in from there. The young sailor did very well on his first long run, in spite of fairly heavy rolling at times. He did his trick at the wheel both morning and afternoon and seemed to thoroughly enjoy it. I opened the wheelhouse windows to let the breeze blow through his hair while I pack him on one arm, pace the deck, and steer with the other hand. He got a mouthful of spray once when we forgot to dodge but didn't mind a bit. I am sorry to say Glenys finally had to give up trying to feed him and for the last week he has been on evaporated milk entirely. Fortunately it agrees. Glenys is picking up nicely and is more like herself again now.

Don't let Dad work too hard on that sidehill.

Feb. 7

Never a dull day, hour, even minute, nowadays. I have never got along with less sleep—and felt better for it. Glenys is doing well, in spite of harder work than she was

Union Steamships' Lady Cecilia *aground at Irvine's Landing, December 24, 1944.*

Columbia Coast Mission hospital in Garden Bay on opening day, 1931.

Showing off our new crewman to his grandfather.

First Narrows, *pastel.*

SPILSBURY

ever used to doing, she is getting steadily stronger and more cheerful. I had the doctor come down to the boat to check things over the other day and he finds Glenys slightly anemic so gave her some iron pills.

The young man is coming along fast. We are almost sorry and feel that pretty soon he won't be a baby any more. For a week now he has been practising smiling and laughing several times a day and is really getting quite good at it. This accomplishment came about quite suddenly after his first serious attempts to make talking noises. Previously he had only two moods—crying or quiet.

Feb. 18

Had Ronnie weighed today and he's 10-2 now so he's coming along ok. Glenys too, is much better and stronger. Took a blood test and her "hob-goblins" are up to 70% now. Were 62.5% when tested in Vancouver so this indicates improvement, we are told.

FIFTEEN

Spilsbury & Hepburn Ltd.

As RADIO TECHNOLOGY PROGRESSED, people's demands changed. At first they were happy to be able to pick up Morse code on a crystal and headphones, then they all got hooked on voice. First it came over a scratchy headset, then a scratchy horn. Later the big console models with full booming sound came along and everybody wanted that. But along with this, there was an awareness developing along the coast that radio could make an even greater step toward eliminating the hardest aspects of isolation, and this was by providing two-way communication.

I was in on the earliest stages of this with my ham station, which I built in 1926, but for many years the Canadian Marconi Company claimed to have an absolute monopoly on all radio-telephone transmitters. If anyone else built one, Marconi would sue them. They claimed to hold all the patents. Finally a guy named Ed Chisholm just started in building and selling transmitters under Marconi's nose. And he got away with it, so that opened the way. He built a very good set for all the packers and canneries of the Canadian Fishing Company, then he was taken on by the BC Forest Service and built equipment for their stations all up the coast. Chisholm was a great bullshitter with a very domineering manner, who fought with everybody. He was one of these people who couldn't say a nice word to anybody. He was all edges. His customers bought from him because the product was good and there was nobody else. He would make them a set and say, "There it is. If it doesn't work, don't come back to me." That was the way he treated them. Once I got a start, it was easy to take customers away from him.

I think the first transmitter I made was for Merrill, Ring & Moore, at their Theodosia Arm camp. I was up there fixing receiver sets one time, and they had old equipment built by somebody long before the war, with a station at Theodosia, a station in Duncan Bay near Campbell River, and one in Vancouver. The Vancouver station was a regular commercial broadcasting station which would interrupt its musical programs and broadcast messages to these camps. Strictly illegal as we know it now, and very clumsy, but they were still using it. The set they had at Theodosia was a big, clumsy, inefficient thing that required them to trek down the hill from the office, crank up a high-voltage gasoline generator, climb the hill, warm the set up, make their call, rush back down the hill to shut the light plant off again, then run back up the stairs again and tune up the receiver to get the reply to their call. They couldn't receive while the generator was running, it made so much interference. They'd been struggling away like this for years. Then the generator would burn out and to have it rewound cost about a thousand dollars each time.

I'd had my ham station working for some time at this point and I thought I could fix them up with one like it pretty easy, so George Moore said, "Go ahead. If you think you can do it, try it." They had lots of money. I built them a little two-tube transmitter, about six inches square, that ran off four B batteries. It worked like a damn. They could send their messages to Vancouver through Cape Lazo. It worked so well one or two other people came along and got similar sets built, then old R.M. (Dick) Andrews came along, bought Twin Islands, and got me to put a big installation in there.

That was funny. Andrews was a multimillionaire who operated a large concern called Andrews and George, with head offices in Denmark and Tokyo. He was a mining engineer who developed a lot of the mines in Korea and got into providing machinery to the Japanese in a very big way. He was right in with the Japanese government, and a year before Pearl Harbour he could see what was going to happen. He got his agent in Vancouver to find him an island to live on until the war was over. They bought Twin Islands, just off the south end of Cortes Island, from old George Macauley, a handlogger. Finally old Andrews packed his toothbrush and beat it back from Japan just weeks before Pearl Harbour. He left everything there. He said there was no way around it, these military men were going to force this war, and he'd always known it. He let a contract with a young guy from Lund, Rick Rasmussen, to build him a fourteen-bedroom log cabin on the neck between the two islands. He was a very trusting old man. He'd meet you once, look you in the eye and either he would trust you or not. Half the time he got stung. This young guy from Lund did a beautiful job, put in power plants, central oil heating, every luxury.

Somebody said I ought to drop by to see if I could sell him something. I put it off until one day I found myself near, and he said, "Yes, I want *a radio in every bedroom.*" I made the biggest sale I ever made in my life. I brought in the best RCA Victor sets, installed them, put up noise-free antennae, concealed all wiring, put a big set in his living room, and then came the question, could I put in radio-telephones? and I said yes. I built one for his boat, the *Twin Isles*, and then one for the house, which was the biggest thing I ever made. Spare no cost. I put most of it together on board the *Five B.R.* at Leo Kent's dock in Cortes Bay. I had Leo building the cabinets in his machine shop and sheet-metal shop as I built the components. It was one of the most elaborate stations on the coast and I was quite nervous about how it would work. What Andrews wanted was something that he could pick up and call his office in Tokyo with and I had to tie it into the local telephone network in Vancouver, which was then the Northwest Telephone Company. When the job was finally done he came in for the big test and handed me a number to call. It went through, and when the answer came it was from his son Bill in Tokyo. Then Andrews passed the handset over to me and said "Go ahead and talk!" I was speechless.

I may have mentioned earlier that one of the regulars on our ham network was a fellow named Jim Hepburn. Hep was a young radio enthusiast who during the war worked as office boy for Island Tug and Barge in Victoria. He was very clever at putting sets together and when the wartime restriction came off marine transmitters, he got the job of building radio-telephones for all the Island Tug and Barge vessels. It was then I got him to build my first radio-telephone, which I installed on the *Five B.R.*, and we began working quite closely together on other projects as well.

For some time I had been struggling with the problem of relaying my messages among the customers, Vancouver, and Savary Island. As I said, I had pieces of paper chasing me all over the coast and it was working less and less well as business grew. I began to see this as the principal limitation on expanding my market, and in my mind I had already made the decision I had been shying away from most of

Early Spilsbury radiophone built for Provincial Police, 1938.

Big radiophone I built for R.M. Andrews' yacht Twin Isles *in 1941.*

Jim Hepburn in his Saturday soldier suit.

my life, namely, to establish a base of some sort in Vancouver. Hepburn was very aware of the problems that were forcing me into this, and as events drew us together we finally accepted the obvious and decided to become partners, with him stationed in Vancouver, where he could direct traffic and build radio-telephone sets full time.

The first mention of it in my letters home occurs on January 30, 1941:

> I phoned Hep, and he got special leave and came over last Monday night. Apparently he had written me some time ago and the letter must be waiting at Pender Harbour. After due and careful consideration he is definitely interested in my proposal. While he was over here we went into the various angles pretty thoroughly and I think he will make up his mind in about six weeks. You can't rush Hep. On the other hand, of course, the move is a drastic one for him, and deserves careful thought. In the meantime he has arranged to work at his old job only half time and in the balance of the time he is going to collaborate with me in building some of this equipment, and has already gone back to Victoria with plans and blueprints for the jobs. In the course of investigations I discussed the proposed arrangement with Bud Lando the lawyer, and then went and saw John Weeden the accountant who gave me a great deal of time and good advice re: bookkeeping etc.

> Feb. 7

> Since my last letter things have been advancing rapidly regarding the business end. A logging camp operator who runs two camps up Seymour Inlet, Sam Dumaresq, came down to the boat the other day to get prices and I now have him on the dotted line for a 50-watt radio-telephone. This makes two going in next April at approximately $750 apiece. Two more are in the balance. Hep is already well advanced with construction of the first one, and I phoned again last night with all details of the next. He says "let 'em come!"

> Since he returned to Victoria Hep has talked things over with various legal friends and as a result drew up a sketch of the basis of the proposed arrangements. In general it

sounded quite good as it agrees very closely with what I have in mind. Roughly it is as follows:

- I incorporate as a private company
- I finance the entire setup, contribute all my present stock, equipment, contracts, goodwill, etc. This will be all assessed and will represent 60% of the co. stock, which I will hold as my controlling share.
- The boat remains my own and I will receive a suitable rental from the co. for its use.
- I draw a salary of say $90 a month.
- Hep draws a salary of $90 a month made up of $70 cash and $20 in co. stock. After he has received his 40% of the stock in this way, he will go on the straight $90 cash salary like myself. He will have the option of buying himself in faster if he desires.
- Books will be balanced after six months to see if our salaries are in line with the co. earnings. They should be kept as large as possible, as co. dividends are subject to 30% income tax, war profits tax, etc. At the end of the year, after setting aside a substantial reserve, profits will be divided in proportion to our holdings.

We drew up the papers and tackled the next problem, which was finding a suitable location in Vancouver. We had to be on the waterfront, and there was really only one place on the Vancouver waterfront in those days where you could hope to be part of the small boat scene — Coal Harbour. Hepburn searched for a building, then just for a vacant piece of land we could throw up a shack on, but got nowhere. I went to town and spent more than two weeks searching and pleading, but the only places available were far beyond our meagre means. We considered looking elsewhere, but I had always tied the *Five B.R.* in Coal Harbour, usually at Menchions Shipyards or the Pacific Coyle Navigation dock, and that was where my customers expected to find me. There was one whole section between Georgia Street and the water with the remains of a burned sawmill on it that turned out to be owned by R.V. Winch. Remembering his ties with my uncle Ben, I took heart and went to see the old man, but he was looking for a customer for the whole parcel and didn't want to let any of it out. I can't blame him—that's where the Bayshore Hotel is now.

In the end it came down to a scrap of land on the east side of Cardero Street where the CPR tracks ended. It was unused, and too small to be saleable. I went to the CPR real estate office on the top floor of the railway station and prostrated myself before a gentleman by the name of Mr. Pretty. He was sympathetic, but assured me the land in question was CPR "trackage" area, and sacred to the use of CP Rail. I left crestfallen, but returned several times with different propositions. I offered to rent it and make our building portable, but Mr. Pretty insisted this was completely out of the question. He didn't even have the authority to touch "trackage" area. If the railway needed to build another spur and found he'd alienated the space, he'd be court-martialled or something. I left again, more depressed than ever, but also more desperate. I went over all the other rejected options once more, rejected them all again, and decided it still had to be the CPR location.

I went down to the boat and took an old faded piece of paper out of my ancient history file, then went to see Mr. Pretty again.

"Do you remember the Great Silver Thaw of 1935?" I asked.

"Do I ever," he said. The company lost a whole passenger train. He'd been up for four days straight. How could he forget it? But why did I ask? I took out the piece of paper, which was the letter Assistant General Manager C.A. Cotterell had sent me in 1935 thanking me for relaying all the CPR messages via ham radio.

"Well, well. So that was you!" Mr. Pretty said.

He'd written the letter for Cotterell. Cotterell was now General Manager and Pretty had a meeting with him the next day.

Two days later we had a lease on the property. It was subject to cancellation on thirty days notice and our building had to be on skids—it couldn't be permanently attached—but the rent was only ten dollars a month. Glenys had inherited a bit of money from her mother so I borrowed $1,500 from her to build a little fifteen-by-twenty-foot shop. In 1987 it was still there, and still on skids.

Early in 1941 I sent a letter out to my customers announcing that we'd incorporated under the name of Spilsbury and Hepburn Ltd.

"The new shop in Vancouver will be completely equipped to handle every phase of the work," I wrote, "and we intend to devote ourselves almost entirely to work from up the coast. The same general policies that have been a part of my business for the past 15 years will be retained...I will continue as before to travel up the coast in the *Five B.R.* and the new arrangement will leave me free to make more frequent trips than has been possible in the past."

It was the right thing to do, and really it worked awfully well. Hepburn was very capable and as long as his enthusiasm held up he was hard to stop. Once we got set up in the new shop we began manufacturing sets to a much higher professional standard than I had been able to on my own. I was able to devote more time to public relations, and with a more marketable product in hand, just as the wartime economy really began to gear up along the coast, business surged forward.

I was very anxious to get things rolling as fast as possible. I sensed we were at a pivotal stage, as I explained to Dad in February:

> Every day we lose means more jobs down the river for us. The licence (radio) year starts April 1st. It wouldn't surprise me to see 20 new stations go in before summer, and we stand to get our share of them. The only alternative is Marconi. Marconi are caught with their pants down, so to speak, as they actually have no suitable equipment to offer for this work at the present time, and before they can get anything that is designed for the work we must have our line firmly established. To try and stall things off, Marconi are going around offering special reductions in return for 3-year rental contracts with the various concerns. Once they get an outfit tied up like this of course, they can dish out any old kind of junk to them. Until I can show that we are established, and ready to give service on our jobs, I can't very well go to the big companies and in the meantime Marconi are putting all the pressure on them they can.
>
> The present plans, accordingly, are as follows — we will try to get as much preparation work done as possible so that when I go north in April to install these new sets I can use the opportunity to make contacts, advertise the new business, and generally lay the groundwork for new sales. This trip will take me right to Seymour Inlet (just past the top end of Vancouver Island). I will have two outfits on board as samples and will make it my business to go into as many camps as possible on the way up. This will include many camps I have never had the opportunity of

calling on before and will open up territory I have long
wanted to investigate.

I had never spent so long in Vancouver in my life. We tied up at
the Coyle Navigation Company dock and I ran back and forth
between the boat and the shop in a frenzy to get things ready. I note
in a letter to Mother and Dad April 14 that Glenys "is very fed up
because I work every night and haven't been out with her at all." I
could chuckle, because I never doubted at that stage it was the most
temporary of problems. And for then it was. We were in the midst of
making ready for our longest ever trip north, and would soon be
seeing more of each other than we could bear.

Living so long at the dock, everything had come unstowed and the
boat began to be very crowded. The radio room was full of
everything from blankets to high-chairs. I had a grand straightening
around to make room, and then stowed the following: two large
console radios, six mantel radios, two large radio-telephones in steel
cabinets with extra speakers, telephone handsets, aerial equipment
and accessories, two Johnson 110-volt combination generating
plants, five extra-large heavy-duty storage batteries, six cases of B
batteries, four cases of assorted batteries, oscilloscopes, analyzer,
wavemeter and tool kits. It represented a couple of truck loads. But
once it was put away there seemed to be more room around the boat
than before.

We passed two days at Halfmoon Bay, then did a job at Vaucroft
on Thormanby Island, arriving in Pender Harbour about 11 P.M. I
sold two top-of-the-line Stromberg-Carlsons there and then
portaged into Sakinaw Lake to answer a request for service up there.
I had been called in by this old couple before and it was a damn
nuisance fighting my way up the lake, but as I explained to Mother
and Dad, "They're English and so don't mind the expense." You'd
wonder how I could say a thing like that after dealing with old
Kendricks and Sutherland, wouldn't you?

After a short stop at Vananda, we put into Savary, where I
installed a small set for Laurencia Herchmer and spent a day helping
Dad with the old Fairbanks Morse light plant. I had some work to
do on a new station going in at a logging camp in Chonat Bay on the
top end of Quadra Island, so we decided to put in at Refuge Cove,
where the Tindalls operated the store. The Chonat job went as
though it was oiled and I had the station running the first day,
making a very good impression on the boss there. Their logging

operation had a serious breakdown and they saved the price of a plane charter by using the radio, which just about paid for it right there. We hopped across Johnstone Strait to Rock Bay at 4 P.M. and just got securely tied up when we had to move out to let a plane come in for a man who'd had both legs broken. The Reverend Greene was there to greet them and managed to bum a plane ride back to town at the same time. What with a good fat nurse, two crew and baggage, the little plane was pretty well loaded but managed to stagger up off the water somehow. Glenys went up to the hospital to get Ronnie weighed but when she got there she took such a dislike to the doctor and his offhand manner she didn't go through with it. I wrote Mother and Dad telling them to forward the next mail to Allison Harbour and we set out for Palmer Bay to spend the night.

The next morning we started for Havannah Channel, figuring to overnight at Cracroft, but we made such good time we decided to keep running right up Johnstone Straight to Alert Bay. The Marshall Wells hardware boat was in there with Daniel Wetmore and his wife, so I was called upon to get out my movie projector and show all our latest stunning chronicles of the BC coast, which were in fact comprised mostly of shots of Ronnie. After that we got practically no sleep, as steamers going by caused such a disturbance that we bumped all night. In the morning the weather looked doubtful, so we didn't make an early start and ended up doing a couple of radio jobs. Then Bob Richardson of the Acme Radio Service, the competitor I had been playing hide-and-seek with the length of the coast for the past five years but had never actually met face to face before, showed up. True to his legend, he was drunk and gushing with good fellowship, and we wasted a lot more time before getting away in the afternoon.

The weather looked better by then, so instead of following up the Island to Hardy Bay we headed directly out across Queen Charlotte Straits. We couldn't make out the shoreline on the other side for about two hours, so I ran a compass course, and when the little bumps on the horizon grew into mountains we were across. The next two hours were up the mainland shore and exposed to the open Pacific. We had never experienced the big rollers before, but they were more fun than frightening once you got used to the idea of moving up and down so much. Actually it was a very fine day and all seemed perfectly smooth to the casual glance, but every once in a while a plume of spray would shoot up like a waterspout, and when it subsided there would be an island. Then another smother of foam, and the island would be gone. We arrived in Allison Harbour in time for supper, the trip having taken six hours.

It was all new to me. I had never been that far north before and there seemed a world of difference in the landscape. I had to sit down and write Dad about it:

> The country around here is quite different to anything around our way. All comparatively low. Dozens of little islands like plum puddings and covered with feathery-looking scrub cedar. Everything thickly wooded, yet no timber worth mentioning. Mostly scrub cedar with a few scrub hemlock thrown in, and every here and there what appears to be a yew tree. Further inland large spruce grow in the creeks and draws, but so far at least have seen *no firs* and it does look funny. Allison Harbour is perfectly landlocked and although it is only 25 minutes run out into the open Pacific it is so calm it is almost stagnant — grass and moss growing all along the boomsticks.
>
> The whole town is on floats including the oil station, warehouse, store, cafe, and numerous residences. Don't think anyone has set foot ashore for years. Actually only one family here anyhow. Pop runs the boat, Ma runs the store, Sis runs the cafe, Sonny runs the errands and so on. We will be calling there on our way out again to collect mail etc. and the old boy is a prospect for a small radio station. The *Venture*, which left town Wednesday, doesn't reach Allison till tonight, 2½ days out of town, and still going north! Whatta trip that must be.

We had a wonderful night's sleep there and made up for what we lost at Alert Bay. We left there at noon the following day, caught the slack through Schooner Pass and the Nakwakto Rapids, and made our way around the many zigzags forming Seymour Inlet, Nugent Sound, Belize Inlet, Mereworth Sound...in the rain they all looked alike. Small mountains, prickly with cedars, no burnt-off country, no sign of logging till you got way back in, and then as far as I could see it was all one-donkey logging, timber right on the beach, and all cedar and spruce. Our destination was the Dumaresq camp, where I had managed to talk the owners into taking this fifty-watt radio-telephone. Up to that time people didn't think radio would work so far from Vancouver, and I wasn't sure myself, but I felt I would have a chance with a high enough aerial. The whole camp was on one float about three hundred feet long, just covered with buildings of all descriptions. The masts had to be raised right in the middle of this muddle so as to give the guy lines sufficient spread. The longest masts they could find were 116-foot *yellow cedar* poles, since there was no Douglas fir in that country. They were so limber they both broke during the raising and had to be spliced, so then they stood only 105 feet high. It wasn't what I wanted, but I prayed under my breath it would pull in enough of a signal to satisfy them.

While a crew of men fought to get the masts solid, I went to work wiring up the set. I stretched the aerial leads out of the office and draped them over the railing beside the walkway ready to be strung up as soon as the masts were ready. Then I went back in to turn on the set. I just wanted to make sure it was ready and didn't expect anything to happen, but to my amazement sound just boomed out of it. Vancouver was coming in like a freight train, and the only aerial it had were the leads draped over the railing! About that time these poor guys who'd been wrestling with the masts all day staggered in covered with sweat and cedar bark and said they were finally ready to hook me up and that there squawk-box sure as hell better work after all that. I hurriedly switched the set off and assured them they didn't have a thing to worry about. I figured the reception would come in even better with a proper high-altitude aerial, but to my embarrassment, it didn't work at all. I pretended I knew what I was doing and proceeded to string another aerial just as low as I could make it, and the set blared forth once more.

What I didn't know at that time was that the height of the aerial didn't matter as much as the angle it formed with the part of the sky that reflected the signal. The signal went up from the transmitter in Vancouver, bounced off the atmospheric stratum radio people called the heavy side layer, and then down to the aerial in Seymour Inlet — at an angle the high-level aerial missed but the low one just nicely intercepted. Another thing I would come to realize as I set up more stations on floatcamps was that the salt water acted as a very powerful reflector, gathering and strengthening the signal when the aerial was placed closer to it. Nobody could tell me any of this because nobody knew. I had to find it all out for myself.

I didn't have the heart to tell the Dumaresqs all the work they'd done on the big masts was for nought, so I left them hooked up, but told the men they needed the low one too, as a backup in case of storms.

"I really like this country," I wrote Dad. "So green and peaceful. Can't get over the way they burn cedar in all the stoves and even the donkeys. As a special treat they burn yellow cedar as it gives a lot more heat than the ordinary. They use it around the cookhouse. Saw it up in the water with a drag saw and it smells to high heaven but they don't seem to notice it any more. They even have to use cedar and spruce to make runners for the donkey sleds."

The Nakwakto Rapids were disappointing after the glowing reports we had heard of them. I daresay they are pretty fast on the big tides, but are very short and straight and only get big runs when there has been a heavy rainfall. We stayed in the inlet for a week, figuring to take the mainland side on our way south so we could do all the main places as far as Minstrel Island.

We pulled out of Allison Harbour after lunch and got a nasty westerly all the way down Queen Charlotte Strait to Wells Passage. The tide was running out of Wells Pass so strong we could hardly buck it, and we stood out in the rip for about two hours trying to make as many miles. We tied up at O'Brien's camp in the Pass for the night and went ashore in the morning to snoop around. I found I knew two of the men there, so we soon got talking. The timekeeper was running the camp and I gave him a great pep-talk. I told him how many sets we had installed in the past few days and how much money radio was saving at Chonat Bay. I was just so eager to make a sale, oh boy. Too eager. I described the situation in a letter to Hep:

> Here is the setup. They need one badly, but it will have to be handled diplomatically. A couple years ago they apparently had a Marconi outfit of their own — cost them $400 a year — but it was so unsatisfactory that when they moved camp they had to throw it out. Now they go up to Earle and Brown's camp at Claydon Bay every time they want to telephone, and the timekeeper says he has gone up there as much as five days in a row and still not been able to get through, and says if radio-telephones aren't any better than that, they don't want some. He says to see the owner, George O'Brien, and see what we can do, but thinks the boss won't bite again. However, if we put them up a reasonable proposition, say 6 months on approval at $25 a month with option to purchase if satisfied, we should be able to do something with them.

There's only one family-owned set here and it's away in

the hands of Richardson at the moment. He spent two days on it in camp last week without being able to find the trouble. It was a set he'd sold them at that. Don't know if I told you — I met Richardson for the first time when we stopped at Alert Bay on the way up. He was half tight all the time, and boat, wife and radios in a filthy mess. Carries no test equipment of any kind beyond a cheap Stark tube tester. Don't think his competition is going to hurt us much.

Our next stop was Earl and Brown's camp in Claydon Bay, at the far end of Wells Pass. There were several camps in here, and a large collection of floating buildings including a store and a freight shed, similar to Simoom Sound. In fact many of these floats were traded back and forth between Simoom, O'Brien Bay, Sullivan Bay, Echo Bay and other floating communities around the Kingcome Inlet area. You never knew what you'd find in any one place until you pulled in. Claydon Bay at this time was one of the more impressive concentrations of floating flotsam and jetsam, with nineteen permanent (or semi-permanent; nothing was permanent in the floatcamp world) families in addition to all the single men in bunkhouses. Of course there were also cookhouses, shops and A-frames. I had high hopes coming in but, strangely, very little work came my way and we would have pulled out except the *Venture* was four days late bringing the latest mail from Hep and we had to wait for it. There was just one repair job, and it was a dilly, as I wrote Hep:

Rather tickled as I just washed my hands after fixing up a Marconi radio that several local experts gave up as hopeless, Richardson wouldn't touch, and came back completely incapacitated after a trip to Marconi's in town. Hadn't run for two years. Some bright egg had soldered a .5 MFD bypass capacitor across aerial and ground on the band change switch in an attempt to improve screen bypass. In addition the secondary of the ANT transformer was open, the primary RF transformer was open, and the electrolytic across the bias supply was shorted due to someone connecting the B batteries up backwards. The threads were stripped on several of the trimmer condensers and it was generally battered up.

After getting it running I pruned 19 turns off all three coils and made it go down to 1630 kHz for the first time in its life. Also found a secret padding condenser which all the other boys had missed and got it working half heartedly on short wave, which was a distinct shock to the owner. When I got through I told them to bring in something tough next time.

I spent a very satisfying but otherwise unprofitable day nursing that old Marconi back into shape. I might not have bothered if we hadn't been stuck waiting for the boat, and I would have left Claydon Bay without having broken through. But what happened next was remarkable. First the Marconi owner came back just oozing pride and ordered a dynamic speaker to replace the old magnetic thing he had. Later he returned again, escorting two other radio owners with problems. By the time the *Venture* finally arrived I was bogged down with work and loaded up with orders for new equipment. I was still there, still working when the boat came back through on its return trip.

Apparently Richardson had made a particularly black name for himself around those parts and the whole community had become very wary of anyone travelling under the name of radio expert. One man had freighted up a frightfully expensive Marconi console model which hadn't worked in over a year, but had chosen to buy two further sets rather than entrust himself to the hands of travelling experts or send the problems to town, where of course the city technicians had no appreciation of upcoast reception problems. The old wrecked Marconi had been given to me as a test, with everybody holding back to see how I fared. I wondered how the first victim had got up nerve to take the plunge and finally he confided in me. He had among his prized possessions a very ancient 201A-type Atwater Kent radio of the binocular coil variety, and glued firmly to the lid of this relic was one of my old service cards bearing the notation "Repaired Nov. 14, 1929." I had given up the practice of gluing cards on sets I repaired because it never seemed to do any good, but I had been too impatient. This old set had been knocking around the coast for twelve years before coming home to roost. And did it ever pay off! Claydon Bay became an absolute goldmine for us, and gave us a foothold in that part of the coast we would build on for years.

Claydon Bay was the highlight of that trip. We finally got out from under the work on May 17 and moved on to O'Brien Bay, then

Gregory Island in Kingcome Inlet. The next jump was to Simoom and Echo Bay, then on down to Minstrel and back into familiar territory.

There was terrific interest taken in Ronnie wherever we went, much to Glenys' despair. We would hardly get tied up at a camp when some woman would run out and say, "Oh, you've got a baby on board, can we come and see him?" From then on they would tell their neighbours and come along in bunches. Glenys tried to tell them he was shy or tired, but as soon as they appeared he perked right up and did his tricks for them...never *ever* folded up when there were strangers around to admire him. Glenys said she felt she should be charging 25 cents for the Red Cross.

Back in town, Hep had been having his hands full trying to get the work on the new shop building wound up and keep up on orders. He had cheques amounting to $1,943 waiting for me to countersign and was positively alarmed at the outflow of cash. So was I, but I felt it was my job to keep up morale and told him not to let it get to him, it was only a lot of marks on paper, and would soon be replaced by a lot of other marks saying we were just as far on the other side of the ledger. It was a bit harder to cheer him up about having his paycheque delayed, being used to a regular salary as he was. "Once business gets under way at the town end it will keep him so durned busy he won't have a chance to wonder," I wrote Dad, half hopefully. I recounted our progress:

> Totted up roughly the other day and find we have a total of something like 19 boats and 9 land stations for whose operation we are responsible. This includes all makes of gear, but about ten of our own construction. Last week landed a job for a new receiver for Vancr. Tugs to be built for $142.50 and a complete transmitter and receiver for General Towing Co. $485 plus installation charges. Several more nibbling but not pushing as have about all we can do for a while.

We spent the rest of the summer of 1941 on the south coast and much of the fall tied up in Vancouver again, getting the shop into shape and trying to get some sets built between phone calls. We took a run up Jervis Inlet in October and made our usual calls at Pender Harbour, Savary, Refuge Cove, and Vananda. I wrote Dad that Ronnie had now taken up waving at people, and waved at everything that moved, even boats on the other side of the channel you could barely see. Tied up at the dock in Pender Harbour waiting for the steamer to blow, I was very excited to discover that he showed vague interest in a photo of himself and set out to determine scientifically whether he really recognized himself, or merely saw it as another toy. I took one of the innumerable close-ups we had of him in his bath and held it in front of him upside-down. He looked at it carefully but showed no great interest. Then I turned it right side up and he immediately broke into a smile of delight just as when he saw himself in the mirror. I tried other pictures and duly reported to Dad, "He seems to be able to recognize himself in almost any pose in a picture." I felt I had made quite a breakthrough in child psychology.

There was a terrific storm on the weekend that drowned a Pender Harbour fisherman setting his net just outside the harbour and wrapped a tow of hog-fuel barges around Merry Island lighthouse. On our way south we ploughed through acres of sawdust and wood chips floating upon Malaspina Strait. We stayed in Vancouver helping Hep catch up on orders for several weeks, then in November we set out again for a long trip as far up as Claydon Bay. Two days in Pender, overnight at Savary, two days at Refuge, then onto Redonda Bay and Stuart Island. At Stuart Island we stopped to mail all our arrival notices for the rest of the trip, but the postmaster, John Willcox, forgot to put them on the steamer and we ended up arriving ahead of them in most places. From Stuart Island we went back down through the Hole-in-the-Wall to Waiatt Bay, Owen Bay and Okisollo, then Thurston Bay. The big Forest Service depot at Thurston Bay had just been moved out, leaving only two houses for the local ranger and his assistant. Rock Bay was the same. The camp was all gone except for a crew of men picking up steel. It had been the biggest logging camp on the coast, built after the Hastings Mill Company had finished taking all the good timber from around Burrard Inlet. On my earlier trips I'd ride their train forty miles inland to where their main camp was. They had four hundred men in there, and I'd fix a dozen radio sets. Thinking what a busy centre it had always been gave me a desolate feeling. But in spite of those two places, it was turning out to be a phenomenally busy trip. I had run out of sets and batteries and had to wait there to meet the steamer to replenish stock.

Ronnie was now on the move. He was getting wilder and harder to handle every day. "The amount of pep is a source of continual amazement to us," I wrote Dad. "An amount quite unmatched by anything else we have come across at his age. We give him a chance to play with someone his own size and he inevitably puts them in a flight of panic. He's a Blitzkreig on all fours!" After turning her back for a moment Glenys would find him sitting on the tray of his highchair with his feet dangling over the edge, just ready to crash down onto the table. At night we strapped him to his mattress. In the middle of the night we'd find him out in the saloon knocking stuff around with his bed strapped to his back like a turtle, gurgling and chirping as if everything were normal. We lived in fear of him somehow managing to fall overboard. We got a leather harness for him and kept him on a rein from then on.

I had a derrick for hoisting the dinghy and I'd put a trolling spring on the hook and hoist him in his harness until his feet were just touching the deck. He'd bounce and bounce and bounce, oh God, I can hear that spring working yet: zing-zing-zing. People would come and watch and he was happier than hell, laughing and chattering away. Someone from around the area who was just a little bit smarter than me patented that idea, named it the Jolly Jumper, and made more money selling it than I ever did selling radios, but Ronnie had a million dollars worth of fun out of it. For a special treat I'd swing the davit out overboard and lower him down on the winch till he could just get his feet in the water and he'd get a great kick out of it. His mother didn't approve of that. Once when we thought he was safely tied up he slipped the harness, and he was up on top of the wheelhouse like a shot. Scared the bejesus out of us. But for the most part he was no more hardship on the boat than he would have been ashore. For washing diapers we had a black enamel pot the size of an overgrown saucepan with a little electric swizzler in the lid. You just stuffed in the laundry, buttoned the lid down, plugged it in, it went ka-chonk, ka-chonk, ka-chonk—it was wonderful. I never saw one like it before or since.

We would cross paths with Bob Richardson once in a while, and when we did we'd find slim pickings—all the easy money gone and all the hard jobs left—so we'd try to keep at opposite ends of the coast. On this trip we ran into him at Port Neville and decided to head directly up to the other end of our territory. We set out for Port Harvey, but the weather was so dirty Glenys squawked and we put back in again. It had been blowing southeast for days so I anchored behind the boom, but during the night it switched to northeast and caught us so hard I thought it would blow the paint off the boat.

The new headquarters of Spilsbury and Hepburn Ltd., 1941.

Ronnie, the "blitzkreig on all fours," temporarily contained in one of my battery boxes.

Union Steamships Chelohsin *docking in Sullivan Bay.*

Railway log dump during the heyday of steam logging.

Gusts knocked the spotlight around, and on shore we could see branches flying through the air like a snowstorm. The alders were bent over flat. We were glad to be out on the water where it was safe. We heard later that several boats sunk in Alert Bay. The next day we managed to get our nose into Havannah Channel and out of danger. From there on up, good shelter is so plentiful you can't get caught like you can in Johnstone Straits. We got into Claydon Bay December 6. We stayed a week and did sixteen jobs, then went up Drury Inlet about halfway, then halfway up Kingcome Inlet. I wrote Dad:

> It really is grand country around here. Very pictur-esque — all big rock bluffs, green trees, and little fluffy looking islands covered with cedar, some no larger than our boat, but always green. Timber is practically all cedar or hemlock of course, and everywhere we tie up there is a strong smell of freshly sawn yellow cedar, and the sharp smelling smoke from the chimneys. Most of them burn yellow cedar exclusively as they get nothing for it if they send it to town in the boom.

We were in Claydon Bay during the bombing of Pearl Harbour. The effect was immediate, even up there. Overnight we had blackouts. The people took the announcement so seriously that when I looked out at 9 P.M. I thought we must have drifted out of the bay. Not a glimmer to be seen. The whole mood of the coast darkened accordingly. Then we *knew* we were at war. Couldn't use running lights. Lighthouses were covered up. Everything was black. They stopped giving weather reports. We were in the midst of violent southeast storms that natives said were the worst of the year and we had to proceed by guess and by God.

SIXTEEN

Pearl Harbour Panic

THE FIVE B.R.'S OLD CLETRAC ENGINE was worn out, so we stopped in Blind Creek and put it in Leo Kent's machine shop for overhaul. We didn't get back to Vancouver till the middle of February. We found the war panic running high on the south coast, with some rather unexpected implications for our old friend Andrews at nearby Twin Islands. All the time he was in Japan, Andrews had been diverting his money to this agent, a prominent Vancouver lawyer, to escape Japanese restrictions on foreigners taking money out of the country, but when he came to collect, the agent looked him in the eye and said, "What money?" Andrews couldn't sue because what he was doing was illegal in an international sense. Then the agent got appointed to some wartime position in Ottawa and inflamed military authorities with the notion that Andrews was a Japanese spy sending reports direct to Tokyo with this high-powered radiophone, and they came up and confiscated all the fancy radio equipment I had built. Roy Allen on the police boat, my great friend and fellow ham, whom I had also built several sets for over the years, was commandeered to come up from Powell River and remove this equipment. He knew it was my pride and joy, and he very carefully wrapped it up, placed it in safe cartons, and handed it over to the army.

Andrews was away in New York but by the time we got into Vancouver he had returned and called me up. "You know, Dinky," he said—he always called me Dinky, which apparently was the

Japanese word for radio technician. Andrews spoke and wrote Japanese fluently, and even taught Japanese writing to the Japanese themselves. "You know, Dinky, you are about the only one in Canada who hasn't taken me to the cleaners. I appreciate it." Everybody else had maimed him. Japanese, Chinese, Koreans he could do business with, but Canadians — look out. "Tell me, Dinky," he said. "Do you think you could do me a favour?" He wanted me to come along to see General A.V. Alexander of the Canadian Army headquarters in Vancouver. I went and explained how this equipment was designed for one purpose only, which was to link up to the BC Telephone Company in Vancouver. It was quite incapable of being used to talk directly to Japan, even if the operator were a magician. General Alexander ended up by apologizing to Andrews for the conduct of his subordinates, and the equipment was turned over to me by some rather sulky army functionaries lower down in the chain of command, but they had been kicking it around and it was in a hell of a mess.

These lower-downs reluctantly permitted me to go and reinstall it but I had to satisfy them that in the case of a Japanese invasion they would be able to destroy the set instantaneously. I installed a switch under the table that if you flipped it one way it was for regular use, but the other way would double the voltage and burn everything out. They came up on a 110-foot launch bristling with armament to inspect it before we could turn the radio back on, and as soon as they had given their approval and disappeared around the point, I disconnected the switch. I didn't want anybody to rush in and throw it on self-destruct by accident. I spent a lot of time on old Andrews, and the funny thing was, I might have been the only guy who gave him a fair break, but I was also just about the only guy who didn't end up with anything to show for it. I wound up with bare wages and that's all. Some of the guys who screwed him got enough to retire on.

Hep had the shop in good shape, with lots of work on hand and the place looking very full and well stocked. He had taken on a sixteen-year-old boy named George to do errands and registered him as an apprentice, and it seemed an excellent move. Our first employee! We needed him. We were behind on production and had to buckle down to a few weeks uninterrupted shop work to catch up. We felt in tune with the mood of the city, which was really gearing up for war work, with shipyards going round the clock. My old aunt Edie was doing Red Cross work and everybody was hustling and bustling in every direction.

When the accountant totted up our 1941 sales they came to $14,000. With this heady discovery we raised our salaries to $120 a month. The shop clearly needed another vehicle capable of packing sets and parts and materials around and we hoped to get a panel delivery or half-ton, but these proved to be scarce and pricey. We settled for a 1941 sedan, which we knocked the back seats out of and made into something resembling a light truck. We were going places, alright!

On March 11, 1942, we all dropped tools to rush over to English Bay for a glimpse of the *Queen Elizabeth I* at anchor. She was several miles offshore and looked as big as Texada Island. She was in for refitting as a troop ship at Yarrows and it was all hush-hush, but everyone was talking about it. The roads were jammed with cars out to see it. My aunt Edie wrote saying my cousin Dick, who worked for the Department of Forestry in Victoria, had been ordered up to Prince George for a few weeks and was afraid to leave his wife and three kids alone in Victoria. The people over there were quite convinced they were going to be bombed at any moment. Hep's dad, an engineer, was put to work on plans for the evacuation of Prince Rupert. People did funny things. One woman of our acquaintance confided she had purchased a diaphragm, which was a newfangled birth control device at that time, to have on hand "for when the Japs come." Just how she planned to put this defence ploy into action I didn't have the courage to ask.

We got one radio telephone set built up ahead of orders, and lo and behold, who should come sniffing around for it but the army. Three men from the signal corps popped out of an enormous brown lorry and strutted in, swagger sticks aswagger and brass buttons agleam. They were very peculiar to deal with. They obviously wanted something, but were maddeningly indefinite as to what it was. It was all too hush-hush for them to say. I had no idea whether they were pleased by what they saw or not, but a few days later some other group came by with a massive heap of requisition papers and packed the set away to Victoria.

We were very proud of ourselves for having cracked the army, but two months later it came back to us. They had installed it in an army tug, the *General Lake*, it turned out, and we were called in for service. Again, everything was so top secret nobody could reveal just what the trouble was, but we did manage to ascertain the radio had never worked once, and they would like this remedied. In fact, they had chosen the most unsuitable outfit possible for what they needed.

They ordered it for 32 volts and the boat was wired for 110. We were told it would be used on ordinary towboat frequencies, but some hamhanded person had tried to alter it for their special RCCS frequencies and made a mess of it. We got it all untangled eventually, and the captain was delighted to have something aboard that actually worked. The boat was so overloaded with electric refrigerators and other appliances that we could see the interference problem was going to become impossible once they were out of the harbour, but they couldn't authorize us to install the necessary silencing condensers. They had their own men for that. Their own men had no idea what to do, so we showed them, but it was no use because they didn't have the authority to requisition the parts, and before they could arrange to get the necessary authority the war would be over, so there was no use trying.

We were fighting the red tape war on other fronts, too. Checkpoints were set up all over the coast to monitor small boat traffic, and anything over ten tons had to be registered in Ottawa. The *Five B.R.* wasn't, and furthermore, I hadn't the basic paperwork to start from. Bob Weld had never given me his builder's certificate. He never had one. Now, according to the authorities, it was too late to get one. I was advised the only way to satisfy the regulations would be to declare that the boat had only been built that year, and hadn't existed for the previous eleven years. I was getting in step with the bureaucrats' way of thinking enough to go for this, but Weld was magistrate of Parksville and felt he couldn't tell a lie. It took months to straighten out. Meanwhile, I couldn't leave the dock because the first checkpoint was under the Lions Gate Bridge at the mouth of the harbour. They'd fine you five hundred dollars. Hell, they'd blow you out of the water. They had an eighteen-pound cannon they were just dying to use. They put a hole through a newly-launched liberty ship from Burrard Shipyards, and up at Yorke Island they sank their own water barge.

Then there was the oil controller. They announced gas rationing in March of 1942 that limited nonessential craft to fifteen hours operation annually, in our case about enough to reach Savary in good weather. I filled every can I owned, then went up to the oil controller, but was unable to get satisfaction. I wrote Toronto for clarification, but in the meantime I had a hundred hours worth of fuel hoarded, so I was good for one long trip. It was also becoming increasingly difficult to obtain material and parts for building our sets. The government had taken over the country's entire stock of metals regardless of who possessed them. I finally found a good stock of sheet metal for cabinets, chassis and panels at Terminal Sheet Metal Works. They wanted to do our work, but they couldn't touch their own steel without first getting permission from the government. We had to write the steel controller in Ottawa, state our requirements, say what it was for, who it was for, and wait to see if they saw fit to issue a release. This, of course, took weeks, and meanwhile the logging companies were hollering for their equipment. For material that had to come from the US it was even more complicated. Each order had to be expressed in terms of its defence value, then the Canadian defence value had to be translated into US defence priorities before the US government would release the stuff for export. Any given radio we made needed both kinds of material. I tried shouting at first, but could immediately see that was useless. Then I got the idea of imitating the manner of the signal corps brass and muttering indefinitely about hush-hush work for the army, strategic industries — coastal shipping, and so on. That worked like magic. Other stores and service shops wanted to know how I was doing it, but I just acted mysterious and indicated it was all hush-hush.

We remained tied up at the dock for all these months and life became rather monotonous aboard the *Five B.R.* Ronnie continued to astonish us with his new Tarzan-of-the-jungle personality. He now developed the trick of squealing at the top of his lungs — an ear-piercing, almost super-audible note that never failed to startle. He'd straighten himself out, puff up his chest like a rooster, and let fly. Its occurrence was unpredictable and frequent, and nothing we could do would prevent or modify it. It was very difficult when we had visitors. Glenys' sister Alison came down with her little boy Denis and we could hardly believe what a gentleman he was. Never seemed to get into anything, and just lay quietly when he was put away in his crib. Ronnie would shout and wrestle and climb out, or jerk the crib across the room and demolish everything within reach. When he was awake, he was never still or quiet for an instant. I kept Dad up to date on his progress.

He climbed up the steering wheel yesterday and pushed the starter button and amazed himself. He was so impressed he didn't dare touch it again. Later he was wandering around the wheelhouse and turned the spotlight switch on. Glenys was in the washroom and heard the click and called out, "Ronnie, what are you doing?" Promptly another click as he turns it off then toddles down to display his innocence. He understands pretty well now. Knows his name, "come here," "stop," "No!" and always says "ta" when given anything. I am dying to see him running around in the sun on a green lawn somewhere. There is no place we can let him go in the world we travel through. He is always perpetually cheerful in the course of his mischief. While Glenys was up town I let him into the radio room as a special treat. He immediately got busy with tools and all the knobs he could reach. Suddenly he hit the jackpot, turned the right knob at the right time, and one of the big consoles burst forth in music. He turned and ran straight for where I was squatting, buried his head in my lap and didn't move until I told him everything was alright again. He didn't go near that radio for the rest of the afternoon. Every day has its quota of surprises.

We had the boat hauled out at Aitken's shipyard and copper-painted for the summer. It blew a cylinder head gasket when we started up and I had to airmail to the States for a new one. I got the engine fixed and got back to our regular berth a week later than we planned. It wasn't much fun being cocked up at an angle on the ways with Ronnie aboard. Just as we finished our first supper at "home" we saw the masts of Roy Allen's police launch go by and trailed him down to Coyle's dock, where he tied up for the night to unload contraband goods confiscated from the homes of Japanese settlers up the coast, who had been evacuated in February. We spent the next couple of days cleaning and loading the boat for our first big trip of 1942 and set out May 1 in the evening. We stayed the night at Snug Cove on Bowen Island, then moved up to Pender the next evening, with stops at Roberts Creek and Halfmoon Bay. After three days catching up on work there we moved up to install a station at Stillwater, then over for a visit to Mother and Dad, then on through our regular route to Refuge Cove, Stuart Island, and Okisollo Channel. We spent a week in Port Neville.

Our next lengthy stop going north was five days putting in a station at a camp in Call Creek (now called Call Inlet). Although it was well into June, it was so cold we had to keep the heater going. It rained continually. Finally the wind changed from southeast to west—but it kept raining down drops as big as grapes. They didn't seem to have a fine-weather wind in that country. Call Creek has steep sides, no beaches, and after days and days of low ceiling became rather oppressive. Its chief claim to fame was being the worst place on the whole coast for teredos. Boomsticks crumbled apart after five months. Logs that had been in the water over eight weeks were not accepted by the mills. I saw a new spruce float they'd built for a donkey that April, and it was only six inches above water. The camp was on floats and they were kept busy rolling new logs under them to keep from sinking. They had already rolled all new logs under that spring, but by the time we arrived they were having to block up the sidewalks between buildings to keep their feet dry. They were talking of moving the camp ashore to get away from the bugs but the only flat place they could find was way up on top of a hill. I had a heck of a time raising aerial masts on their soggy floats.

While we were there we usually had lunch and supper at the cookhouse. We didn't want to, but they were so insistent we finally gave in. Ronnie was a problem at first, but later we took his highchair in and he was good as gold, and quite interested in the spectacle of the loggers attacking their grub. He would have his little bit of vegetable and gravy and bread and butter and fruit, and ate very well. Before we left he came to know the sound of the dinner gong and would run to the door as soon as he heard it. I had to walk ashore along a boomstick carrying Ronnie under one arm and the chair in the other, with Glenys hanging on behind. She was not at all steady walking on logs and quite hated it. After we were finished there we moved on to Minstrel Island and down Clio Channel. I had it in mind to explore some new territory at the mouth of Knight Inlet this trip—all the little islands: Village, Crease, Berry, Alder, Midsummer.

It proved interesting travelling, although there was little business. I was fascinated by the Indian villages, which looked a good deal more Indian than what I was used to seeing down south. There were totem poles in abundance and old-style burial houses. In one place a logging company had to move sixteen burial houses off to one side of a logging road. The Department of Indian Affairs was there supervising, although the Indians themselves appeared to take little interest. There were hundreds of tree burials. In one place we passed without going ashore you could see a cedar box perched across two limbs about sixty feet up a spruce tree, with all the other limbs cut off to stubs about three feet long. Some places were absolutely

Echo Bay, Gilford Island.

Study of whirlpool, pastel.   SPILSBURY

Mount Stephens, O'Brien Bay, Tribune Channel, *pastel 1986*   SPILSBURY

Glenys with Ronnie on leash at Village Island Indian village.

Burial house, Village Island.

Grave markers, Village Island.

Carved houseposts, Village Island.

littered with dead Indians. Cedar boards and skulls were scattered through the bush. Fallers at one of the camps said you couldn't stick an axe in a tree without a bunch of bones coming down on your head. I pointed out one tree with three burials in it to Glenys. We decided it must be a family tree. Many were quite recent. The loggers said one of the trees had a silver plaque on it naming some chief and bearing the date 1930. There was a patch on the beach where they apparently burned the deceased's possessions at the funeral, just a heap of melted bottles, china, trunk hinges, rubble. It was a part of the ritual they still performed. I had been hoping to take some photographs but the weather was bad.

From there we went up to Claydon Bay and did several days hard work before heading up Drury Inlet and Acteon Sound. At the head of Acteon Sound there is a lagoon about four miles long, Tsibass Lagoon, that very nearly backs around into the tail end of Seymour Inlet. The entrance to this place, one of several smaller rapids in that area referred to as the "Roaring Hole," is about twenty feet wide and navigable only at high slack. The rest of the time it is just a waterfall. We didn't attempt to enter, as there was only one camp up the lagoon and it would mean a 24-hour layover.

Word came that Hep had a set ready to go into the Canadian Forest Products camp at Englewood on Vancouver Island, so we ran straight down to Englewood, arriving July 4th. I was pleased to discover the boomman was an old friend, Bill Giroux. I'd known him previously at Rock Bay, where he'd also been head boomman. He used to look after my boat while I went inland to the camp, and offered to do the same again here. This was a relief to me because I knew I was going to have to leave Glenys alone on board while I spent several days up Nimpkish Lake installing the station, and I had been worrying about it. If I'd known how long I'd actually be I would have worried more.

This had been a sale set up by Gordon Armstrong, the old ham who now worked as salesman for a company called MacKenzie, White and Dunsmuir. Gordon first got his ham licence in 1919, seven years before me. First his call was VE5UP but then when they changed the numbers for the Vancouver district he became 7UP. He was one of the first people I met on the ham network, and it was him who bought Dad's old Caterpillar tractor. He was another Salesman Sam, and he had worked hard at persuading me to set up in Vancouver, where I could be more useful to him, and now he was turning a considerable amount of repair and installation work our way. But like a lot of salesman types, Gordie wasn't too careful about details, and somewhere along the way the specs for this set had got all mixed up. We ended up with a 100-volt AC set for a camp that was wired for 100-volt DC. This was a big new camp being set up in the territory vacated when Wood and English went broke. I described it to Dad:

> It is about 30 miles back to where the transmitter will be installed temporarily, then in a few months we will have to move it to a new camp site on Woss Lake, another 25 miles or so I believe. This will be the permanent headquarters. They are laying it out like a regular town. Expect to have about 120 families living there besides 600 men. They have over four *billion* feet of timber in their holdings already, with lots more to come. They expect to log at the rate of 20 million feet a month with a top speed of a million a day and at this rate anticipate about 75 years steady logging! This is much the biggest thing of its kind ever to be laid out and the last *big* virgin stand on the continent. Forestry officials plan on logging it in such a way that the young growth will be coming in as fast as they log it. For this purpose they will log it in alternate strips so that the standing timber will seed the logged-off patches.
>
> The railway will be built to last indefinitely, like a main line of the CPR. Another big outfit in the valley is Pacific Mills, the same company that owns Ocean Falls. They have several billion feet to come out, but they are going to use trucks and are now engaged in building 10 miles of *concrete* roadbed 40 ft. wide. They say it is cheaper than planked road, as the trucks last longer and go faster!

I got on the phone to Hep but he had ten jobs on the bench and had just received word that we had 24 hours to strip all our radio equipment off R.M. Andrews' big yacht, the *Twin Isles*, before the Navy took it over. He had just fired our apprentice George for being lazy, and he wasn't in a mood to have me tell him to go back to work on the Englewood set. I didn't want to take the *Five B.R.* all the way into Vancouver and back on our skimpy fuel supplies, so I decided to leave Glenys and Ronnie and dash into town on the Union Steamships' *Catala*. I hadn't ridden on the Union boats since the outbreak of war, and it was an eye-opener. I wrote Dad:

She is loaded to the gunwales with people of all descriptions, mostly military from Prince Rupert. I boarded her last night at Englewood, was not surprised to find there were no berths, so I parked crossways on a leather-covered chair till 6 A.M. this morning—scared to go to the toilet in case I lost the seat—by 5:30 just discovered how to sleep in it and then the lights came on and people started stirring. Woke up very hungry so went right down and bought a breakfast ticket then had to wait for third sitting at 10 A.M. They feed the northern passengers first. By the time the last gong went I was so hungry I went back to the office and traded my 50 cent ticket in on a 75 cent one... At least half the passengers seem to be in uniform—Air Force, Army, and Navy all represented. Some of them were pretty tight last night and didn't settle down till about 4 A.M.

I didn't arrive in time to be any help dismantling the big radio in the *Twin Isles*, so Hep enlisted a couple of plumbers and wrecked the thing right. They just chopped the cables with an axe and threw the pieces in a box. It was all they had time for. Then the navy changed its mind and decided they didn't need the boat after all. They sent it back up to Twin Islands with the dismembered radio in a heap, where it was to await my arrival. Hep was swamped with work. Kirkland, the Englewood boss, was already talking about five more sets up at Nimpkish; Boeing Aircraft was asking about getting radios for the bombers they were going to build in Vancouver; and Edward Lipsett, the fishing supply company, was sending over more and more installation and service work for the US-built Kaar Engineering sets they were selling to the fishing fleet. Gordie Armstrong had now moved his act over to Lipsett's, taking the Kaar agency with him. Marconi had never really broken into the fishing industry, but with Lipsett behind them, Kaar now made great gains. It was a good set for the time, and it overshadowed our own make, but Gordie had arranged to have us do all Lipsett's service and installation work, so we were busier than ever.

I got the adjustments done to the Englewood set and did something I'd never done before: I availed myself of the regular Canadian Pacific Airlines service to Alert Bay and made my first airplane trip up the coast. It was a lovely flight in a Dragon Rapide. I was fascinated to discover how mixed up I could become trying to sort out the coastline I knew so well, and saw a lot of lakes and valleys hiding just back of the shoreline I'd never suspected before. I landed right in front of the *Five B.R.* an hour and forty minutes out of town and gave Glenys the surprise of her life. I was two days ahead of time. As I told Dad, I wasn't sure we were making any money but "we sure see life."

In the meantime it looks like I won't get through here for another two weeks. I went up to camp A yesterday to see just what I'd need. I find they haven't even built a house for the power plant yet. I stirred things up plenty and have a gang building it now. When I got to the beach had to go down to the mill and get a flat car loaded with five thousand feet of lumber and shot up there. Even had to pick out the necessary lumber and tell them how to build it. Thought I had left the building business ten years ago but apparently have to bulldoze things to get anything done around here. Tremendous display of disorganization but that is to be expected in any new camp I guess, particularly one of this size. They are paying us $15 a day as long as I am on the job so I suppose I shouldn't lose any sleep over the length of time specially as I hear they intend spending six million dollars before starting to put logs in the water!

The trip up the railway to the lake was fifteen miles and quite scenic in spots, with high trestle bridges. Nimpkish Lake is fourteen miles long, one and a half wide, and deep. It was just full of steelhead in those days, and I was very impressed at the swells it could produce in a brisk westerly. And the blinking thing was full of seals. I counted twenty in one herd the boat passed through. One of them was snow white. I never would have believed they could survive in fresh water, but the men told me they stayed year round, feeding on the trout. There were also wild swans nesting on the lake. We saw six of them, together with a grey one they claimed was a "swoose." I spent a day on the boat getting things ready and listening to Ronnie practise his new vocabulary. He had decided to be practical and start out with "bread" and "milk" but you wouldn't have known it to listen to him. The resemblance was plain only to him, but it was the same noise every time and we got to know.

On Sunday we went over to Alert Bay for groceries and found the CNR steamer *Prince Robert*, now rigged out as a cruiser, in port for a big blowout with the Lieutenant Governor aboard. Everybody in town was drunk, even the customs officer, who was supposed to clear us in and out of port, and we had to leave without papers, risking a penalty too horrible to think about. Word came that Hep had been ordered to take a medical exam and had been classed C-1. This meant he would probably be conscripted, although not immediately. Like me, he had volunteered at the beginning of the war and had been turned down flat in spite of everything he could do to get in. It was a wild card that added to the various pressures upon us. The Englewood job continued to make absolutely no progress. I ended up taking charge of installing the generators and wiring the camp, none of which was my affair, but even then I couldn't get co-operation from the various mechanics and machinists who were supposed to supply the parts.

I seemed to spend most of my time travelling back and forth on a wide assortment of vehicles. While I was waiting around on one occasion I hitched a speeder ride about three miles out a branch line to what they called Camp 9. This was an old Wood and English camp that I reckoned must have been abandoned about three years, judging by the calendars left on the bunkhouse walls and two-foot-high alders growing up between the ties. The fireweed and thistles were four feet high, and set up a great cloud of down and leaves behind the speeder. Everything had been left as they were using it—the utensils in the kitchen, the oil in the lamps. The cookhouse had sixteen tables set for twelve men each, so it was no small camp. There was a terrific amount of machinery and parts left piled around in confusion, including some switches and insulators we needed back at the lake.

I asked the speederman how much further the track went. He had no idea but offered to take a run out and see, if I wasn't in a hurry. Naturally, I agreed. We started quite slowly, thinking to run out of track every few yards, but eventually we got our nerve up and were clipping along about thirty miles an hour. The track was in good condition except for the odd place where gravel had run down from a cut. We beetled along up this unnamed valley, which ran parallel to the Nimpkish Valley but about five miles east, passing through cuts and over trestles by the dozen. We passed a small lake about half a mile in length, and several dry lake bottoms grown over with thick grass through which were trails, probably bear trails to judge by the abundance of sign. Finally it cut through a depression and into another valley. It followed up the side of this for about a mile or so and then made a U-turn and came back along the other side, gradually climbing to an altitude of about a thousand feet I should think.

It was too cloudy to see the sun and neither of us had any idea which way we were travelling by now. We could see around quite well, as the whole country was logged off to the last stick, leaving an endless plain of stumps and fireweed. After hours of this we came to abandoned equipment, including three donkeys and several cars on sidings. If clouds hadn't been laying down so low I believe we should have been able to see out into the Straits again, as I believe this second valley was taking us back out that way. Then the track turned abruptly and headed back into the island again up a third valley. We quit at this point because we didn't want to run out of gas so far from base.

I spent the next few weeks asking anybody I could find about the country we'd seen, trying to determine just how much further we could have gone, but no one knew, and few showed any interest. I couldn't get over it. It was like a lost world out there, a vast ruin as big as a European country, which had been created so casually it had no name and nobody cared it was there. Still, there were ghosts about the place. I couldn't stop trying to picture those loggers, logging out their days in that nameless valley and then vanishing with such mysterious finality their works were forgotten by the time the alder was two feet high.

Later, up at Camp A, away up at the head of the lake, I found more abandoned locomotives and donkeys and machine shops and piles and piles of steel, but these were much older workings, probably abandoned twelve or more years before. In one place there was a coal dump with probably a hundred tons of coal right behind the camp, but you couldn't see it because the alder trees growing in it were about six inches through. In all there were 35 donkeys and 12 steam locomotives scattered around in those woods, remnants of the Wood and English empire just left to waste. You couldn't help wondering about the meaning of it. What had driven them to do it, to create this barrenland so great it was probably big enough to be seen from the moon? Money? But the whole enterprise had ended in bankruptcy, so what did that mean? Whatever it meant, it wasn't giving any pause to this new wave of loggers busily setting up among the ruins, eager to swipe the last great fir forest off the face of the earth. Nor to me, just as eagerly providing their need for communication. We were caught in something, we heirs of James

Ward, that drove us to gouge canals, flatten forests, manufacture radios; something we were driven to do as much in spite of the end result as for it. There were moments like this that gave one pause, as the whole industrial process seemed to resolve itself into a questionable pattern, but then the picture wavered and soon one found oneself back pitching in with the best of them.

SEVENTEEN

War Work

WE WERE TIED UP TO THAT BOOMSTICK at Englewood for over a month. It was hard on Glenys, being stuck on the boat and having to take the dinghy ashore for exercise or supplies. Ronnie was becoming quite used to the life, however. I wrote Dad:

> He was off his grub for a bit a couple of weeks ago but now he is definitely back on it again and eating everything in sight. Very noisy and lively. Makes a serious attempt at talking now. Says "bree" which means either bread, butter, or breakfast. Also "boah" for boat and "buh" for bird. If you mention bread to him just once he goes to the cupboard, opens the door, drags out the loaf, then pulls out the knife drawer and gets stopped before he can get hold of the breadknife.
>
> Well, the steamer is in so must go ashore and mail this. Heaps of love from all. Ronnie just heard us say we were going ashore so he turns up with his lifebelt pulled over his head but no pants on!

It was August 5 before we finally got the last details cleared up at the camp. When we came to untie, Bill Giroux asked us if we wouldn't like to take that boomstick with us! It was too late in the day to do any serious running but we were so eager to see the end of Englewood we struck out anyway, making Port Neville. The next day was a beauty and the Hansen's old log homestead, with its little bit of lawn and weeping willow over the pump, looked so enticing we couldn't resist going ashore. Ronnie had a marvelous time running around with the two other kids, the dog and the three cats. We lay around on the lawn and Mrs. Hansen served coffee and cake. It was about the only relaxation we got that summer. I was so late getting back I skipped practically all our other stops and headed directly in to town. Hep was itching to get away on his holidays and there was a station waiting to go in at the new airport the Air Force was building at Fort Rupert, as well as three big Kaar sets from Lipsett's waiting to be installed in fishboats. I had to run up on the *Chelosin* to do the Fort Rupert job, and then batten myself down in the shop until Hep came back. I had a heck of a time trying to do anything while the phone was ringing and people were dropping by. I'd try to get back up to do a little catchup work after supper, but Glenys was always hounding me to come home earlier. I wrote Dad:

> Things have stacked up pretty well and we will have a very busy week or ten days catching up on everything. While Hep was away I sold a $590 set for the tug *Atlas* and that has to be installed next week. Also have one of the Kaar sets to install on a brand new fishboat, the *Pacific Girl*, just launched at Menchions. This is next week too. Then Kelley Logging sent their transmitter in from their camp in the Queen Charlottes for overhaul and this has to be shipped out in time to catch the bimonthly steamer at the end of next week, so I hope Hep is in good shape after his holidays.

The *Five B.R.* stayed tied up pretty well until Christmas. We were still having problems with the oil controller and by the time I was caught up on the work it was well into fall. The storms were dreadful that year. I have never seen it blow like that in Vancouver and I was fascinated watching the city bear up under a real gale. Going over the Cambie Bridge with the car in the afternoon, I was afraid we were going to be blown off it. People were hanging on to poles and railings to keep from being blown away. In the middle of it the power went off and streetcars all over town stopped in their tracks. Driving as we were, we couldn't figure it out. Why were all these cars stopped at weird angles going around corners or halfway along blocks? Were the Japs coming or something? Some people were trying to get off, others trying to get on. It must have lasted half an hour or more. At Granville and Georgia there was a streetcar parked crossways under the light and traffic was tied up as far as you could see.

We hired a new helper, a radio amateur who was working part time at Point Grey Wireless and needed something to fill in. He proved very useful, since work continued to build.

In November I took the car over to the Island on the CPR steamer *Elaine* and drove out to Port Alberni to do an installation at the Kildonan cannery in Alberni Canal. I had to take the local boat, a seventy-foot diesel called the *Uchuck*. It was packed. I watched a company of soldiers, 61 including officers, march aboard three abreast complete with tin hats, bayonets, packs and luggage, and disappear somewhere below. In addition there must have been 25 civilians and possibly 20 individual soldiers travelling on their own. I noticed that the ship's papers allowed her to carry a maximum of 34 passengers. The trip took about three hours and I was darned glad to get off. It was snowing most of the time and it was quite impossible to get inside as the cabins were full of women and kids. Kildonan was on the right-hand side of the canal going out and consisted of nothing but the cannery and associated buildings. It was one of the largest on the coast at that time, populated, true to its name, largely by Scotch people. On the way back through Parksville I stopped in to see my old ham buddy 5BL, Bob Weld, who was full of talk about the civilian aircraft detection corps, first aid, and everything else. He had been an artillery major in World War I with service in the Mediterranean, and he had a fairly low opinion of anything military.

Without exactly planning it we were finding ourselves spending longer and longer stretches pinned down in Vancouver and had begun to speculate idly on the usefulness of getting a little bungalow of some sort ashore. Then we made the mistake of actually starting to drive around looking at places and before we knew it we were completely caught up in house-searching. We made a deposit on one place, but missed it. We were heartbroken. We took up the search then a little less frantically, having a hard time getting very interested in anything else we saw. We were getting to the point of giving up when I decided to have one last stab and went through every ad in the Saturday *Province*. We borrowed the company car and made the rounds of 22 places in a couple of hours, discarding them as fast as we came to them. House is in a hole—roof looks in bad shape—faces wrong way—too close to the streetcar—too far from the streetcar — no view — looks too old — can't stand pink stucco and red facings — rest of the houses in the street look too dilapidated — rest of them look too prosperous and make this one look like a cold potato served up with lemon pie — it went on and on. Just before we quit for good we found a little place on the hill above 16th Avenue in the Dunbar area with five rooms and only seven years old. We bought it for five thousand dollars.

We had been coming to rely more and more on the Kaar business coming over from Lipsett's, and it worried me. They had become the most popular set around, and this was due in no small part to our installing them and keeping them running. We had built up the Kaar business at the expense of our own make and we had nothing but verbal assurances Lipsetts would cut us in on the benefits. Then the thing I dreaded happened: they brought in a new manager and he decided we were making too much money. They set up their own service department and it left us with damn near nothing to do. But in December the war priorities people caught up with Lipsett's and stopped them importing Kaar, which was made in Oregon. I jumped in and talked up an order for ten of our own sets to fill their outstanding orders. This was much our largest deal to date and we celebrated by going out and hiring a part-time bookkeeper and receptionist, Mrs. Norah Yates. Within a year Lipsett's were out of the business, leaving it once again to us and Marconi.

A week before Christmas we got away for a short trip up the coast, stopping in to spend Christmas day with the Tindalls in Refuge Cove. We followed our usual route through Stuart Island, Surge Narrows, Chonat Bay, Rock Bay, and back down to Savary before stopping off January 19 at Powell River to go up and work on a set at Gordon Pascha Lake. The morning of the 20th it was snowing heavily and I thought the trip might be called off, but the camp foreman, Tom Murphy, turned up about ten o'clock with chains on his car and we started out. Some parts of the road had about five inches while in other areas practically none had fallen. We had lunch in the cookhouse on the beach at Stillwater before heading up the lake. It was still snowing so we decided to get the speeder instead of taking the car up the lake. The speeder wouldn't work. The snow was too deep for it. Next we went for the gas locie but this was backed off into a siding and we couldn't get it out. Finally we went into the big roundhouse and got up steam on the big steam locie. This took about an hour. With the engineer, the brakeman, the beach foreman acting as fireman, Murphy and myself all squeezed into the cab, we went up the hill with a great clanging of bells and snorting of steam. At the lake we shifted into a waiting tug and kept moving north. Visibility was poor on account of the snow, but the lake was free of ice. When we neared the upper end we encountered a very stiff northeast wind. It must have been forty miles an hour and it got so rough the little tug had all it wanted. We couldn't land at the regular float on account of the seas and had to scramble ashore across a lot of drift logs and rickety planks coated with ice.

The snow was fine and powdery and drove along before the wind like flour. It had drifted into deep banks against the buildings. I thought I would lose both my ears and my nose before we got inside. My job on the radio didn't take long and I spent the rest of the afternoon and evening reading through old magazines and hugging the big drum heater in the office. That night I slept cold with eight blankets on me. In the morning I felt like a steamroller had gone over me.

The next morning it was still blowing as we beat our way down the lake to the waiting locie and puffed and wiggled and clanked down the hill again. It was quite amusing to watch them operate the thing. Everything was built so heavy and clumsy. Even the little tray to set the oilcan on was made of boiler plate. To blow the whistle there was a nifty arrangement consisting of a three-foot piece of two-by-four hinged from the ceiling of the cab and connected to a system of heavy bell-cranks, cast brackets, and half-inch rod. The engineer used enough effort to stop the train every time he blew the whistle. They didn't seem to mind and sat around shouting at each other, never looking where they were going, being the only engine in the territory.

I got back to the boat in time to clear customs at Powell River in preparation for an early start the next morning. We had a squeaky, tumbly night that was cold even with both Iolanthe heaters going full throttle, and in the morning the boat was covered with saltwater ice. There was an ice moustache around the bow and the lines were all festooned with it. The wind seemed to cut right through the hull. The wind was still fresh and there was a good chop, but I decided to pull out in spite of protests and it calmed considerably as we went south. Then as we came up on Jervis Inlet the wind freshened and a snow storm approached from the east. It would be dark before we got to Pender Harbour and I didn't like the looks of it, so I decided to put into the little cove behind Hardy Island Roy Allen always spoke of as being the perfect shelter. I guess it was for either southeast or westerly winds, but this had to be the first true northeasterly for several years and it wasn't worth a darn. The wind funnelled straight down Jervis at us and the boat bumped violently all night. The bumpers were rock-like with ice and the ropes so icy I couldn't loosen the knots to take up slack, so we were swinging in cornerwise and hammering against the float. Out in the centre of the bay it was all whitecaps and so cold the sea had a four-foot ice fog over it.

But we were lucky. On the radio we learned the government float broke away at Stuart Island and smashed up the landing. They radioed for a tug and the *Polar Bear* was as near as Mermaid Bay, but with the wind pouring out of Bute at sixty miles an hour he didn't dare risk the ten-minute crossing. The *Prospective* was close too, but he said he already had five inches of ice on him and he was afraid he would ice up worse if he went out. The Powell River tugs lost two Davis rafts off the north of Vancouver Island and there were several scows lost. We heard on the radio news of the Frank Waterhouse steamer *Northolm* sinking off the west coast with only two survivors.

The morning turned out bright and sunny in spite of the cold wind, so we decided to take Ronnie ashore to see Hardy Island's famous tame deer. The island was kept as an estate by a wealthy Californian named McComber whom nobody ever saw, but the caretaker, an incredible old character named Tom Brazil, was known to everybody on the coast. He had tamed these deer just for the fun of it, and had as many as 75 at times, though there were only 30 at this point. They were very friendly and let you fondle them. Ronnie got a big kick out of feeding them apples. Old Tom said they came and went as they pleased but always came back. They lived an average of twelve years, according to him, with a few running to fifteen. Some of the bucks were quite handsome. I tried to get movies, but it was so cold I couldn't handle the camera very well, and Ronnie was quite inconsolable in spite of his snowsuit and mittens. The thermometer on the verandah stood at sixteen degrees Fahrenheit, but it was the north wind that gave it the bite. The old boy made us come into his cabin to warm up, but more importantly, so he could show us his photograph collection. It was mostly pictures that people had taken of the deer, some of them many years before. He also had a series on his tame bear, which he'd raised from a cub until it was so big he could ride it. The bear got to like it, and he'd ride it out hunting. One time though, he was out on a long trip and it got late. When he tried to get on the bear it started acting up on him and he had to give it quite a beating with a fir limb before it settled down and let him get on. He was pretty upset until he got home and found his own bear there ahead of him. The one he was riding was a wild one he'd got hold of by mistake. This was a typical Tom Brazil yarn. He told them one after the other as long as you could stand to listen.

The wind kept blowing, so after lunch we got up our nerve and set out anyway. When we got around Cape Cockburn we found a heavy deadswell coming down from the northwest, with the wind blowing offshore from the northeast and setting up a rough chop, so the travelling was not bad at all. Pender Harbour had practically no breeze and the sun was quite warming. The people there didn't even seem to know it had been blowing in the rest of the world.

We spent our usual four days in Pender catching up on business and then back to the city just in time to save our new house before all the plumbing froze and split. We began the long-avoided task of moving off our boat, the only home we'd ever known together, in February. The main move took nine carloads, piled to the ceiling each time. I had in the meantime arranged for my good old 1931 Plymouth to be shipped down from Powell River on the steamer, so now I was firmly established in suburban life, with both house and auto. I wrote Dad enthusiastically about the quickness of the drive from 28th and Dunbar to the foot of Cardero, which in those days of gentle traffic I could accomplish in four and a half minutes. My letters became increasingly concerned with tips for the proper treatment of brown spots in the lawn and the pruning of hedges, which Dad answered as if he had been waiting all his life for me to ask.

I made only one more service trip in the boat, the following winter, and then let it sit. There were a number of reasons. The demands of business were such now that I really couldn't justify being out of town and out of touch for months at a time. Cardero Street, not the *Five B.R.*, had become the centre where major decisions and deals had to be made, and as the senior partner, I had reason to be on hand more than the system of serving the coast by prolonged boat trips allowed. But more pressingly, there was the oil controller. We had never clearly been granted exempt status under gas rationing and had to get by on temporary rations. As the war progressed, these became more difficult to obtain. In the meantime, I found myself travelling more frequently on the passenger steamer or the plane.

EIGHTEEN

Robert B. Gayer

THE MOST MEMORABLE TRIP THAT SPRING was one of those unplanned things that just popped up. One day this guy walked in and asked to see the manager. Mrs. Yates had to handle such situations very carefully, as to whether she should call Hepburn or me. She would size the customer up and decide in her own mind which one of us would be the best to handle the particular situation. We were both back in the shop working at the bench. She called me this time. This guy's name was Robert B. Gayer, and he told such an unlikely-sounding story that I decided to call Hep in so he could hear it for himself. This was something so big and important that I felt it required both of us. The responsibility would be too much for one.

The way Gayer explained it, he was the director and general manager of a large mining concern on Vancouver Island, CANgold Mining and Exploration Company Limited. They had an operation on top of a five-thousand-foot mountain up near the head of Great Central Lake, and they simply had to have good communication with the outside world. This of course meant *radio-telephone*, and only the best would do. "None of these half-assed things you press a button to talk and then listen to the other guy and you can't stop him," says Gayer. "We already have one of those goddamn things, and I wouldn't give it hell room. I want full two-way telephones and to any part of the goddamn world, any time of the night and day! Time is money goddamn it in my business and I won't fart around with anything less. Now, can you do it?"

We took a deep breath and assured him this was all entirely possible and to give us an hour or so and we would work out a cost for him.

After he left, Hep and I compared notes. In the first place we realized that this little "good-for-nothing" radio he had been deriding was actually one of our own make—our little AD-10 transportable, and we hoped he wouldn't notice the name on the box before we sold him a new one. I told Hep I didn't like the look of this guy. There was something about him. He wore expensive clothes, fancy shoes, his hair was slicked back, he had a smooth face and a pencil-line moustache. He looked like a Mississippi gambler. I wouldn't trust him as far as I could throw him, and judging by his weight, that wouldn't be very far. Hep didn't agree. He said, "I know that type. I've seen lots of them in the towboat owners' group. Don't let appearances fool you. I bet he's worth stacks of money!" Anyway we compromised. I said to tell him that he required a very special type of set we would have to build specially, and ask him for a cheque in advance, then cash the cheque before we started.

That's the way it was. All but the installation, and that was something else again.

When the set was ready, I was to go to the mine site and put it in. We contacted Gayer. He said, "No problem. Go to Nanaimo, through Parksville towards Alberni, and take the Great Central Lake turnoff. I'll meet you at the lake and take you from there." Just that simple. At Great Central Lake Gayer met me with a small tug and half a dozen of his men. We chugged up to the head of the lake (it's nearly forty miles) and unloaded all our gear on the beach. Then the fun started. For the rest of the trip Gayer had a whole bunch of men and about twenty pack horses. The radio-telephone, generating plant, and batteries, along with tons of general supplies, were loaded aboard these wretched horses. I had expected to be met by a bus or station wagon, but about now I was beginning to suspect that this would not be so. Finally Gayer said, "Here's your mount. Jump aboard!"

Well, I think the last time I was that close to a horse was Bob Tipton's on Read Island, when we used it to drag John Jones' new radio through the woods. I hadn't actually sat on top of one since a brief pony ride when I was about five years old. When Gayer wasn't looking I got the thing alongside a stump and climbed up that way. Fortunately I didn't have to steer because all the horses just followed each other like a chain. There was just this one narrow trail which switched back and forth up the mountain. We were soon in snow. We were under big timber all the way. When we reached the two-thousand-foot elevation the snow measured 28 feet deep. They had apparently broken the trail with snowshoes.

I was a bit upset at one point. Two of the horses slipped over the edge and rolled down the hill with their packs. Gayer calmly took his 30-30 carbine out of his saddle holster and shot both horses, then had the men remove their loads and put them on the other horses. He explained that both had broken their legs and there was no other recourse. He said the gullies around there were full of dead horses.

We finally reached a spot where everybody stopped and dismounted. There were some blazes on the trees that indicated that we had arrived at the main campsite. Then they started to dig—and dig. I have a photo of this—28 feet down, at the bottom of the hole, they encountered the roof of the camp building. It was a split cedar shake roof. They pried off about a dozen shakes and there you could look down into the main building. We all climbed down into total darkness and a loud, musty smell, and there we were. They lit gas lanterns and started a fire in the wood stove, and by this means created enough draft to draw some fresh air down the entrance tunnel. After a camp supper we climbed into our sleeping bags for a peaceful night under the snow.

In the morning I got busy and installed the radio-telephone. I got someone to climb a couple of trees and erected a fairly respectable antenna system, but I had to route the antenna feeders down the tunnel and through the hole in the roof to the set. Then I had to set up the gasoline-driven generating plant. This couldn't be in the house because there was no way to get rid of the exhaust fumes. Gayer came to my rescue and built a little platform in the fork of a tree close to the cabin. I bolted the plant down to this platform and ran the electric wires down the snow tunnel through the hole in the roof and to the set. Now all they had to do when they wanted to talk was climb up the tunnel, go over to the tree, start the engine, and go down the hole again and do their talking. Incidentally, it worked very well indeed, and Gayer was very pleased with the whole thing, which was a great relief to me—I didn't fancy running back to this place every few days on warranty calls!

But there was an amusing aftermath. Many months later, Gayer sent me a photo of the generator plant in the tree. In this picture it was forty feet above ground, and crosspieces had been spiked to the tree all the way down. Apparently every time the snow level melted down another couple of feet, they would nail on another rung. No one thought to move the generator.

While speaking of the unusual, there was another feature of the camp that attracted my attention. Toilet facilities were very simple. The privy was at the end of a short covered walk to a cliff edge where it had a free fall of about forty feet. Good enough to satisfy the most fastidious — you'd think. But the space from seat level down the cliff face for forty feet was entirely boarded up like a well casing. I asked Gayer about this, wondering if it was to cope with high crosswinds or something. I was sorry I had mentioned it, Gayer was so upset about it. He said that the previous year he had agreed to the foreman bringing his wife up to live in camp. She was a very modest lady and was not about to expose herself to anyone who might be coming up the mountain at that time, so hubby got a couple of carpenters to box it in all the way down. Gayer moaned that it cost him 26 cents a pound just to have the lumber hauled up to the site from the lake.

NINETEEN

I Discover the West Coast

I WAS HOME FROM THE GAYER escapade only a few days before setting out on another, longer trip, to install a big government station at Winter Harbour near the northern end of Vancouver Island's west coast. I left Vancouver on the CPR *Princess Elizabeth*, having checked my bags and all twelve crates of gear through to Winter Harbour via the *Princess Maquinna*. It was another wonderful day, presenting a different but equally impressive view of Mount Baker, but everyone's attention was captured by another, more immediate attraction, as RCAF planes streaked continually overhead. Some were just circling out from Lulu Island, some were having bombing practice on the target range near Saltspring. Others went over in a beeline as if they had someplace to go. When we got to Active Pass, a Spitfire discovered us, dived and roared by at four hundred miles an hour right at deck level about thirty feet away. After circling he came at us again, this time from dead ahead, and just zipped over the masts with no more than a few feet to spare. The speed of these things was so great that you could hardly twist your head fast enough to follow them. He repeated the performance several times and then was joined by a second Spitfire who did it all over again on his own. Finally something else attracted their attention and they zipped over the hills.

The next act in the show was a couple of twin-engined bombers. Their trick was to come at the boat sideways, flying about ten feet off the water and holding their course until they were about 150 feet away; then, when you figured they had to crash, they would hop over the boat, practically going between the masts. I imagine their speed was around two hundred miles an hour. They were medium bombers, with machine guns and about six crew. Some passengers who had come over on the earlier boat said they had been at it all day. They were from Pat Bay airfield.

I had a couple of hours layover in Victoria and used it to call on Mother's friend Leigh (Burpee) Robertson and her husband Robbie. Robbie was now just recovering from a cataract operation and nearly blind, but they were delighted to exchange family gossip. Robbie had some years earlier encouraged my interest in art by teaching me pastel technique and introducing me to Emily Carr, whom I thought a revolting old crank. God, what a messy woman! I never have learned to like her painting, but on this trip Leigh gave me her little book *Klee Wyck* which I read as I travelled north through some of the places she described, and I was very impressed. The coast comes through in her writing, whereas in her painting it is somehow muffled. To this day I feel she was a better writer than painter.

The *Maquinna* pulled out about midnight and I turned in. When I woke up we were unloading at Port Renfrew. I saw practically every stop from then on, as the ship tied up during darkness due to the blackout. I hadn't been to the west coast of Vancouver Island before and enjoyed the sights.

At Clo-oose we dropped our hook about half a mile out and lay in the gently rolling swells while about six dugouts were pulled down the beach and launched into the big surf. They paddled back and forth making several trips each, taking passengers and goods ashore. Transferring the freight to the canoes alongside was quite breathtaking, as the dugout would rise and fall a good ten feet, always managing to miss the ship's plunging guardrail by a hair. Even passengers were unloaded this way. One rather portly Indian lady missed the canoe and was dunked twice before she was unhooked. At another point they unloaded cows right into the sea and they swam ashore with no further assistance. We spent a night unloading in Port Alberni and a whole day in Ucluelet unloading stuff for the Air Force. Ucluelet was in full swing as a big seaplane base while a new field for land planes was being constructed between Ucluelet and Tofino, just behind Long Beach.

At one place we saw a big Russian freighter high and dry on the shore with the surf breaking over her and going at times as high as the masts. On the shore opposite we could see the tent in which the crew took shelter. There had been no news of this disaster because she was loaded with strategic war materials and the wreck was kept under guard. It was generally conceded that she was the victim of our government's singular policy of extinguishing navigational lights, an iniative which thus accomplished the Japanese objective better than they could manage on their own. Not long after this the authorities decided to turn the lighthouses back on again, and kept them on for the duration of the war. We saw historic little Friendly Cove, where Cook and Quadra had their big doings, and the west coast's Refuge Cove, where the Tindalls' mail always got sent by mistake, since renamed Hotsprings Cove to avoid that very problem. We could see steam curling up from the hotsprings on the way in, just twenty feet above tidewater.

Altogether the coast was very much as I had expected it, fairly low-lying, very green, very jagged, with a lot of sand beaches and sand dunes on the hills, especially around Cape Cook. Reefs and rocks lying well off shore seemed to be everywhere, sending the surf up in geysers. In many places the *Maquinna* would thread her way between rocks and passes barely a hundred feet wide. There were a great many day marks and whistle buoys that worked by the swells, continually groaning and sounding very unhappy.

The *Maquinna* was a wonderful boat. Good passenger accommodation, yet an enormous capacity for freight. Three holds and three sets of derricks. A beautiful sea boat, taking everything very easy—no vibration—and yet making a good twelve knots. By this time she had been on the run continuously for thirty years without mishap. During the war she had been fitted with a naval gun on the stern, and two naval gunners kept a continuous watch for subs, not that they could have done a darn thing about them. At the time they were saying there had already been seven torpedoings on the west coast, although I have never seen it verified.

I hadn't done so little or eaten so much in a long time as I did on that trip. All I did was sit in a deck chair and get a tan. I never missed a meal call or a course in any meal. I had the saloon steward trained to pass everything down to my end and then bring me crackers and Stilton cheese to finish up.

Winter Harbour proved to be just what its name implied—open water all around but a landlocked harbour and sand bottom ideal for small boats. It had one store, one fish scow, one oil dock. There were possibly a dozen families in the place, all good typical coast people with good boats making big money. All seemed to get along, and all were enthusiastic about the new radio station, which was lucky because I needed their help. I found I had to raise three masts to take care of the various aerials, one 48 feet, one 60 feet and one 91 feet. It was difficult to find suitable trees. Cedar was scarce and what there was grew scrubby and short. It took me eight and a half days to finish the installation and train them to use it. The night I came we saw sundogs in the west, and the next day it blew as hard as it had all winter and kept it up for six days. It picked saltwater up in twisters, and boats came streaming in from surrounding waters to seek the harbour's shelter. There were half a dozen halibut boats, seine boats and a log barge with tug. Putting the poles up in these conditions was a nightmare. I had to work in the thickest salal I had ever seen: six feet high and loaded with water. It was raining buckets all the time, and blowing right at us horizontally. I got soaked to the skin every day. I finally borrowed an ankle-length rubber coat, but still got wet.

I had a gang of six and they went at it with a will. Although they were very capable people they felt the precariousness of their station keenly and appreciated what radio would do both for their own safety and their ability to help others. They said if the radio had been installed earlier the crew of the *Northolm* could have been saved. If the lookout on Cape Scott had been able to get word to them instead of going to Port Alice, where no one could reach the scene until the next day, they could have reached the crew in four hours, before the men perished.

They did funny things in the way of fishing, to my eye. The halibut season opened while I was there, and these birds would go out on the banks and troll for them. Put their gear down about fifty fathoms and up would come halibut. One man brought in four hundred pounds for the day. They also caught sockeye on the spoon, as well as dog salmon and humpback, something inside fishermen didn't learn about for years. They claimed the dogs and humps were really beautiful fish when caught that way, every bit as nice to eat as a spring salmon. The spring they caught were shipped to Australia, canned and called *ocean trolled salmon*. They had to be perfect, with no scales missing, but they brought a much higher price than inside fish. This type of fishing was still relatively new, having been made possible only with the coming of wire lines and power gurdies.

I was all ready to go, with my stuff all loaded on a fish boat, when

My bags waiting beside the tracks at Englewood.

On the way to the CANgold mine above Great Central Lake.

*View of Tofino
through my porthole
on the Maquinna.*

*Dad in front of
the Port Neville store.*

they had to put through a call, and darned if the radio didn't break down completely from two different causes. I had to unpack my tools and replace a blown tube and a fractured crystal. It was embarrassing, but not so much as if it had happened after I'd got to Vancouver and the government had to bring me all the way back.

I finally arrived home on April 19. I'd only been away fifteen days but I seemed to see a world of difference in Ronnie, imagining he was making great leaps and bounds in his vocabulary. His best one was "buble*bee*" which he'd repeat over and over excitedly and point when ever he saw any sort of insect near a flower. I took this as a sure sign of latent genius until Glenys pointed out that he was well behind Denis and his other playmates in language skills. This was a comedown for me, but Ronnie was happy and busy and obviously didn't care.

We had another installation waiting up on the Nass but I managed to have Hep do it in exchange for my staying to oversee finishing the new addition to our building. We had taken on another helper, Herb Hope, earlier in the year and now had a second full-time helper in the shop, which made a staff of five including Mrs. Yates, and we were forced to build an addition the same size as the original structure to give us a little elbow room. I was painting the inside walls one afternoon in July when who should show up at the door, four hours early for dinner, but Jimmy Anderson.

"Gosh, what an awful job," he said. My face was white, paint was trickling down the inside of my shirt and icicles of the stuff were hanging out of my nose from breathing the spray gun's fog. I had been starving but felt too painted up to venture out, so the first thing I did was send Jim up to the barbecue to buy me something. He returned with four hamburgers and four bottles of ginger ale. He had just finished lunch himself but endeavored to help me with the hamburgers. When we were done I made him take his suit off and strip to his shorts. Then I gave him a pair of coveralls and a paper bag for a hat, and we set to work. We took it turn about and between us we used up eight gallons of the darn stuff. It was something like calcimine but waterproof. The whole time, we were regaling each other with anecdotes of the past and recalling all the queer happenings we could remember, shouting at the tops of our voices over the noise of the compressor and laughing and slapping our knees between jokes. In a little while I slipped up to the little store and brought back two bottles of chocolate milk on ice, two bottles of orange crush, and a pint of ice cream, all of which we dispatched with gusto. It was very hot. Now and then Jim would take a trip over to Menchions' for a pail of water and he presented such a queer spectacle in his baggy coveralls and tall, white paper bag on his head like a chef's hat that I noticed people turning around for a second look, but he was blithely unconscious of it all. We went out later that evening, but we didn't enjoy ourselves nearly so well as we did painting the shop and yelling all day. In fact, when I look back over a long life, that hot afternoon of calcimine, hamburgers, cold drinks and laughter stands out as one of the times I would like to relive.

Ah, life is funny.

Nothing is like it's supposed to be.

In mid-October I left Glenys in Vancouver and went on the last trip in the *Five B.R.*, installing sets at Forward Harbour, Village Island, Acteon Sound, Twin Islands, Teakerne Arm, and Jervis Inlet as well as doing a lot of small jobs and making courtesy calls all over the coast. As many times before, I tried to get Mother and Dad to come with me, knowing their long-held ambition to see the coast, but once again they were "too busy."

That was it.

That made it a lifetime of being too busy, because I was never able to offer them a trip again.

When I got back I'd been out seven weeks and had grown a beard. I had planned to come around to the door dressed as a hobo to see if I could fool Glenys, but she got wind of it and ordered me to leave my beard on the boat. I did.

This was the longest I had been out of Ronnie's sight since he was born and I was a little worried he might not remember who I was, but he made a huge fuss over me. When asked whose boy he was he always replied promptly, "Daddy's boy!" Glenys didn't think I'd notice any difference in him but his talking had improved enormously. He was still very hard to understand but he spoke in sentences, and had taken to inventing his own vocabulary in quite a resourceful way. An axe was a "cut-hammer" and baloney slices "meat wheels."

After that boat trip the oil controller lowered the boom on us and we weren't able to obtain gas in sufficient quantity to make a meaningful trip in the *Five B.R.* We were grounded. Oddly enough, Canadian Marconi chose this exact time to enter the field with a boat of their own. I suppose we could have fought the ruling, or done like a lot of the bootleggers and floating hawkshaws and disguised

ourselves as a fishboat — fishboats were exempt from gas rationing —
but I bought an airplane instead.

TWENTY

We Take to the Air

I HAVE BEEN KNOWN TO SAY that I became involved in the flying business purely by accident. Looking back, it often seemed that way, and in one sense it was true: I never expected what happened to happen. I thought having an airplane of our own would be a good business proposition one way or another. I saw first hand how effortlessly a plane could eliminate those endless boat trips that always stood in the way of anything you wanted to do when you lived or worked on the BC coast. It would open up our radio business. It would keep us a step ahead of Canadian Marconi. And we could probably fill in with a little taxi work around the edges. When we'd had the only car on Savary, people were always coming to us, and having one of the only available seaplanes on the coast would probably be the same. But that was it. I wasn't one of these people like Grant MacConnachie or Russ Baker, who were obsessed with building an aviation empire. On the other hand, I had an interest in planes. If I hadn't, my cousin Rupert would never have been able to talk me into getting that first one.

You will remember Uncle Ben, the uncle who sold the Whonnock property to Dad and went broke with R.V. Winch. Well, old Winch picked himself up and started up all over again, and he sent Uncle Ben a message to the effect that if he came back out, Winch would make him back all the money he'd lost, just like that. Ben came out with his two sons, Dick and Rupert, and my aunt Edie.

Rupert, who was a few years younger than I, took flying lessons down at the old airport at Minoru Park in Richmond, and through him I had a feeling of being connected with the flying world. I would have loved to go down and learn to fly, but knew I couldn't afford to, and I envied Rupert very much. When the war came along Rupert found himself very much in demand, first on the prairies as an instructor, then as a Transatlantic Ferry pilot. He spent a couple of years pushing all the different kinds of planes that were being manufactured and repaired in North America across the ocean for combat service. He was making a lot of money and drinking very heavily and I'd hear weird and wonderful things about him. Once he was supposed to have flown a Lancaster bomber to England without knowing it. He was so drunk he couldn't remember.

Rupert then got out of Ferry Command and came back to Vancouver as test pilot for Canadian Pacific overhaul. "Lord, has he changed," I wrote Dad in February of 1943. "First glance I thought he was Uncle Ben. Very fat in the face and must weigh close to 200 lbs...Mrs. Rupert seems an extremely nice party. Looks quite young and very attractive although she has a daughter of eight." As Canadian Pacific repaired various aircraft, all the way from Spitfires to B-29s, he'd take them up, shake them around, see if the wings came off, and sign them out. It didn't bother him at all. He was a very good pilot. Everyone agreed on that. He could take an airplane by the face and make it do what it was meant to do just by instinct.

One of the things that seemed to go with being a really serious flyer in those days was to be a sort of aviation evangelist, always trying to spread the word about the new frontiers to be conquered by aircraft. The kind of thing Grant McConachie was famous for. Well, Rupert had a touch of this, and between taking his junkpiles aloft he'd come over and get in our way at the shop. Listening to us bemoan the fact we couldn't get up the coast, it wasn't long before he said, "Why don't you get an airplane?"

"An airplane?" I said. "I've barely been in one."

"Sure," he said, "you can get 'em for a dime a dozen now, they've all been grounded by the war."

He made it sound as though we couldn't afford not to do it. He said we could get a little four-place machine on floats that would handle a pilot, three people, take a thousand-pound payload, go 100–120 miles an hour, and only burn 3½ gallons of gas an hour. As a matter of fact, the idea wasn't entirely new to me. Riding along the coast in airplanes the few times I had got me thinking how much easier our work would be if we had a plane at our disposal full time, but I'd never had occasion to consider the feasibility of it seriously. Rupert's suggestion it could all be managed so simply hit a responsive chord. I first broached the subject to Dad in a letter of February 3, 1943. I had been called out to repair the set at Englewood and had a hell of a time getting passage on the CPR. "All of this makes me more determined than ever to get our own plane," I wrote. On March 19 I informed him we had gone as far as applying to Ottawa for permits to operate a plane, and he began to express a bit of concern. I replied putting with my best rationale forward:

> Don't let the idea of a private plane scare you. The idea is in no way new to me but has been in my plans for a good many years now. The time for using one is now practically upon us and as soon as possible we will put it into effect. The use of planes on this coast will become widespread during the next three to five years when they will become as commonplace as gasboats are now. Naturally we don't want to be the last to adopt this method of travel and by starting ahead of the game we will boost our business prestige no end. Just as in the use of a boat, their operation need involve no risk if done with common sense and I anticipate no trouble in this connection. However we have yet to receive a favourable report from Ottawa and then find equipment within our means.

Most of this confident prophecy I was taking on faith from Rupert, but it turned out to be truer than I thought. The scheme took on a much more meaningful air when the oil controller shut down the *Five B.R.* If we could really get a plane, now was the time. So where do we buy one? Well, Rupert took care of that. He found an airplane advertised for sale in Montreal by a man named Albert Racicot, a bush flyer closed down by the war, and he allowed he wanted $2,500 for this aircraft, a Waco Standard four-place biplane with Jacobs Radial L4MB engine. It was serviceable, in good shape, had a valid C of A (Certificate of Airworthiness), and wheels, skis and floats. All complete. Fly it away. But *twenty-five hundred dollars!* This was what I'd paid for the *Five B.R.* and I'd just got that paid off. Well, it turned out that Racicot was a heck of a lot hungrier than it appeared, and I got him down to the point where he said okay, he'd take $500 down, the balance over two years, and he would deliver the machine in Vancouver himself.

That part wasn't too bad, but now what about fuel—how could we get it for an aircraft when we couldn't get it for a boat? How could we get permission to fly when everyone else in the country was grounded? I had now become quite expert at saying the things the war administration liked to hear, and I was able to convince them the small amount of fuel the airplane would use was more than justified by the service we would be able to provide their chain of Aircraft Detection Corps radios on guard against Japanese air strikes up and down the coast. On January 10 I took my first regular business trip by air, leaving Sea Island at 10 A.M. to fix a station at Tipella up Harrison Lake.

It was a clear cold day, practically cloudless. Ponds were frozen over and the upper surfaces of the Waco's wings had white frost. We had to wait for this to thaw off with the engine running, as it cuts flying speed badly. Rupert took off up river and climbed to one thousand feet. By sticking to the river we were able to avoid climbing any higher. This was the first time I'd taken the leisure to really observe the view in an airplane and I recorded my first-time impressions excitedly for Dad:

> Lulu Island and the flat country looks pretty from up top. All fields joined neatly together like a chess board and looking very tidy and varying in colour from greens through browns to red according to growth. Toward Langley there is a crazy pattern of snaky looking-sloughs and creeks winding back and forth in the most aimless way. Roads in very severe straight lines. The smallest details show. The yellow line down the centre of the pavement and the washboard effect of the gravel. Point Roberts, Boundary Bay etc. look very much like a map and ridiculously close even at a thousand feet.

Rupert nudged me and pointed down to the left. Two planes were crossing the river towards us. They were two Kittyhawk fighters painted the usual brown and green and hard to pick out against the ground when they were over it. At first they didn't appear to be moving very fast, then they separated and one came straight toward us, gaining in velocity until it seemed to shoot past in a blur two hundred feet off. It looked like a knife blade coming edge-on, and it was gone before I could open my mouth to exclaim. They circled around and came back past us several times. Up till then our 100 mph had been giving me the sensation of breakneck speed, but their 340 mph left me feeling as if we were practically frozen in place and not going anywhere.

But almost before I realized it, we were over Haney and passing the old family homestead at Whonnock. I was very surprised at the number of farms on top of the hill behind my old home town; hundreds whose existence I had never suspected, revealed to me for the first time by my new vantage point. I watched a long CPR freight which seemed to remain stationary as long as it stayed in sight, although I knew it must be clipping along at top speed. At Agassiz we turned sharply and went in over Harrison Hot Springs. All the sloughs were frozen but the main river was open. The mountains around Hope looked splendid in their crisp white. Harrison Lake looked very green at any time, but from here it seemed unnatural, like a pot of paint. The low-lying snow made the contrast the more garish. The surface of the lake looked a little rippled and we had a discussion on how big the waves were and which way they were going. Rupert thought six inches. I was suspicious, as there were whitecaps showing. We found out a little later they were just about big enough to swamp a rowboat.

It was the last time the look of things from the air would seem quite so strange and marvelous to me. I was dying to take pictures but I didn't have a camera. You weren't allowed to have them in a plane in 1943. You weren't even allowed to carry one onto the airfield.

We taxied up to the slip in a smart wind and I had the job done in about fifteen minutes. No sooner had we got the station on the air than a call came through from Vancouver—for me. It was the camp boss, who wanted to know if we could take a passenger. A man at their other camp had come down with pneumonia and had to be taken to hospital immediately. It took only four minutes to get down to the other camp. Coming back we had the wind with us and made better time. The trip was over a hundred miles each way. We took 1.05 hours up and .55 back. Glenys had waited lunch for me. A trip which had previously taken two days, completed between 10 A.M. and lunch—with time for a mercy flight on the side.

The next week we flew out to the west coast, running across to Parksville and on to Qualicum, then through the pass to Alberni at 2,500 feet. It was a lovely flight. We stopped at the cannery in

My cousin Rupert Spilsbury in Sprott-Shaw's trainer in Richmond.

The age of the radio boat comes to an end: the Five B.R. *up Jervis Inlet.*

The age of the radio plane begins: servicing a seiner on the west coast.

Assembly line at Spilsbury and Tindall.

Ucluelet to check up on Herb Hope, who was doing an installation there, and then went on up to Chamiss Bay, the oldest of the various logging camps run by the Gibson brothers. As usual, it was entirely floating, but it was a world different from any other floatcamp I'd seen. The bunkhouses had one man to a room, varnished floors, carpets, oiled knotty cedar walls, and panelled ceilings. The cookhouse looked more like a restaurant dining room, with regular chairs and tables seating four to a table and a nice view through curtained windows. I did a rush job getting the aerial up, tuning up and putting in the test call two and a half hours after arrival. Then two more installations, one at their spruce camp at Goukinish Inlet and another at Tahsis Canal with a stayover at Kyuquot, a little fishing community much like Winter Harbour.

We ended up taking off on our southern leg at 6:15 P.M. Evening fog was starting to gather in wisps like necklaces around the mountains and looked very decorative from above. The whole coastline was very pretty—all hills well covered with timber and emerald green, lakes deep blue, beaches white. The only indications of activity on the sea were the masses of white foam visible for miles along the shoreline and around every rock. We had to overnight at Zeballos because it was too late to complete the trip into town. It was the first time I'd seen it in daylight. It looked just as one would expect for an upstart mining town, complete with saloons, painted storefronts and a dusty main street, but it was already a ghost town by that time. Four hotels, all empty. Stores all closed. Most of the windows boarded up. We stayed at the Zeballos Hotel, the others being called romantic things like the Privateer and the Golden Gate but none of them very romantic to sleep in.

In the morning the fog was down to a hundred feet and we had to wait until eleven o'clock before we were allowed to proceed. We went down the coast dodging cloud and fog formations till we got to the Muchalat River valley and then headed inland. As most of the fog was along the western seaboard we were soon out of it and climbed up through scattered cloud to five thousand feet and clear sky. Snow-capped mountains on both sides, winding river and small lakes far below. In a very short time we were over logging works and this turned out to be the Campbell River Timber workings coming in from the east coast of Vancouver Island. We came out over Buttle Lake behind Campbell River and headed for Comox, maintaining our altitude. Johnstone Straits from there looked like a river, and in it I could see a fair-sized vessel in the area of Seymour Narrows. I found out later it was the government survey vessel *Stewart*, hard aground on Ripple Rock. The gulf was very smoky and I could hardly see Savary. The new concrete airstrip at Comox was very striking with its gleaming three-mile runways like someone had made a giant "X marks the spot." About twenty huge bombers were lined up on it along with countless interceptors, probably Kittyhawks. I saw one medium bomber taxi out the runway and take off. We were then a mile directly above him. About three minutes later the same plane whizzed past us from behind, circled around apparently reading our numbers, and then dropped down to the field again.

While fuelling up in Comox I put in a call to the shop and found Hep was wanting to see me about a set up Jervis Inlet, where he was travelling with the boat; about twenty minutes later there we were, drinking beer on the *Five B.R.*

Thursday morning we left again. Our first call was Hardwicke Island to dump a passenger. Then back to Blind Channel. Then in to Western Logging at Forward Harbour to fix a station there. After lunch we proceeded to Acteon Sound to fix the station and have supper. We spent the night at the Dumaresq camp in Mereworth Sound, Seymour Inlet. I worked on the station while Rupert caught a cutthroat off the float. On the way home we stopped in at Village Island in Knight Inlet to adjust the station there.

Flying continually was proving very tiring but it certainly got us around and got the work done. We had been used to setting aside a large chunk of our lives to make that trip to Seymour Inlet, and here we were doing it overnight.

I couldn't get over how small the world looked out the window of the Waco. In a single glance I could take in a section of coast that I had known previously only in glimpses stretched over days and weeks of slow boat travel. It was thrilling to be able to look down and see the places where I'd spent my life as mere details in the grand plan, but it was also vaguely unsettling. I knew I would never be able to look at that coastal world in quite the same way. It had become less mysterious, less forbidding, less grand. It really *had* become smaller. That pocket-sized thing out the window was the new coast, and soon everybody would see it. The immense backdrop my life had hitherto been acted out against was abruptly jerked away, leaving me with a slight case of vertigo.

But I knew I was among the first to see the coast's imminent future in this way, and since it couldn't be avoided, I determined to make the most of it. It gave us a tremendous advantage in the radio

Winnifred Hope, manager of Spilsbury and Tindall in 1954.

The coast made small: looking across Desolation Sound from the air with Twin Islands in the foreground.

Jim Spilsbury in 1987.

business, and I could see it opened up possibilities far beyond our own needs. I plunged the company into charter work, selling air travel to the logging companies and mining companies and canning companies. It was one of those moments of historic change when the man who has the right skills and the right ideas is suddenly swept away by events and succeeds beyond his wildest dreams. By 1946 I was president of the third largest airline in Canada. I was lord and master of three hundred employees. I turned over more money each month than James Ward did in a lifetime. After all those years struggling along hand to mouth, boom, it just happened overnight. And in an altogether new field I knew practically nothing about. It was the most exciting time of my life, and much too involved to detail here. I will tell the unlikely story of my "accidental airline" in a later book.

I was completely caught up, but it was a whole different world from the one I'd known. Our planes made daily contact with every bay and island from the south arm of the Fraser to Alaska, but I saw it only in glimpses as I sped on my way to high-pressure meetings in Ottawa and Montreal and London. I saw scarcely more of Glenys and Ronnie or our new children, David and Marie, as I dashed back and forth trying to keep a pioneer airline together without benefit of capital or experience, fighting off one crisis after another. I would be at home most evenings but, as Glenys complained to Dad, I "never got my mind off work," and was nearly always on the phone talking to either Bud Lando or Jack Tindall, who had replaced Hepburn in the radio company, now renamed Spilsbury and Tindall.

First Mother, then Dad, left the old place on Savary and moved to rooms in North Vancouver. In March 1947 came an especially difficult hurdle to pass as I sold the *Five B.R.* I wrote Dad:

> It has been quite a wrench as you can imagine and I think Glenys is more upset than she cares to let on, but I am sure it is for the best. It seems to cut a definite tie with the past, but as I pointed out to Glenys, it is not the act of selling the boat that does that. Our circumstances have changed to such an extent that the old life is not possible, and as far as the boat is concerned, if we didn't sell it, it would fall apart as we watched it.

In 1955 I sold the airline and returned to the radio business full time, putting some of the airline money into an old warehouse on

Cordova Street. Here we built a real factory and became serious manufacturers. The kind of rugged, low-power radio equipment we had learned to build for the small boaters and isolated settlers of the BC coast carried the Spilsbury and Tindall name to every corner of the world. Literally. We sold radios to Turkish border guards, settlers in the Mato Grosso, a French mountaineering expedition in the Himalayas, and the Japanese explorer Naomi Uemura on his one-man polar expedition. By the mid-sixties we were Canada's leading exporter of radiotelephone equipment. In the 1980s the company finally wrote an end to our long running battle with Canadian Marconi's telecommunications division by taking it over. The company carries on today as Spilsbury Communications Ltd.

I retired in 1982 and spend as much time as possible cruising over my old radio route in a fifty-foot diesel cruiser called the *Blithe Spirit*. I was right in seeing the upcoast world on the verge of a great change, but the form the change has taken is still a puzzle to me. I assumed that improving communication and transportation on the coast would bring new growth to the camps and villages and make it an easier place to live. Instead it seems to have opened an easy way for all the residents to escape and do their work from remote headquarters in the city. The world I knew has more than changed, it has all but vanished. I find myself visiting silent, deserted bays carpeted to the waterline with second-growth fir, trying to recall the way they looked when they were laced with railways and dotted with floatcamps and loud with the voices of eager young men and women convinced they had a place to win and a future to possess.

If I turn off the radio and ignore the buzz of the floatplanes overhead, I can almost hear them still.

Index

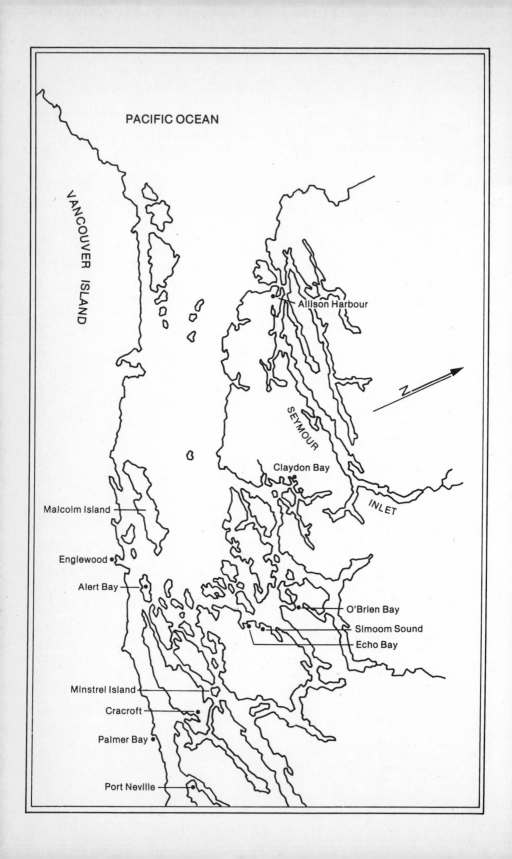

PACIFIC OCEAN

VANCOUVER ISLAND

Allison Harbour

SEYMOUR

Claydon Bay

INLET

Malcolm Island

Englewood

Alert Bay

O'Brien Bay

Simoom Sound

Echo Bay

Minstrel Island

Cracroft

Palmer Bay

Port Neville